Introduction to Fluid Mechanics

RUSSELL W. HENKE

Professor of Mechanical Engineering
Director, Fluid Power Institute
Milwaukee School of Engineering

ADDISON-WESLEY PUBLISHING COMPANY

Reading, Massachusetts · Menlo Park, California
London · Amsterdam · Don Mills, Ontario · Sydney

Third printing — October 1972

 Printed in the United States of America. Published simultaneously in Canada. Library of Congress Catalog Card No. 66-25603.

ISBN 0-201-02809-3
IJKLMNOPQR-AL-8987654321

Preface

This book is a *deliberate departure* from conventional texts on fluid mechanics! Historically, fluid mechanics at first had been the province of the civil engineering community. Later it was joined by the field of mechanical engineering, which dealt with the process industries, power plants, materials transfer, etc., and finally it expanded to the compressible fluid fraternity, whose chief concern was aerodynamics, aerospace, jet propulsion, etc.

It stands to reason, then, that early text material on fluid mechanics was heavily oriented toward hydrology, civil engineering applications, water works, sewerage systems, and flow in open channels. At this stage of development fluid mechanics was synonomous with "hydraulics." And hydraulics meant civil engineering hydraulics. To this day, introductory fluid mechanics courses in most engineering colleges are either taught by the civil engineering department or followed by patterns established by them.

With the evolution of the aircraft industry and its related technology, a second school of thought evolved around compressible fluid mechanics in conjunction with problems of aircraft in flight or aerodynamics. Compressible fluid mechanics, which had been the province of the thermodynamicists, moved into the field of fluid mechanics and became of equal moment with the earlier "hydraulics."

We now had two major schools of thought in fluid mechanics: (1) hydraulics, which encompassed the area of incompressible fluids or liquids, and (2) compressible fluids or gases, which included aerodynamics but did not necessarily restrict itself thereto. Things have been going along this way for some time now.

It is inevitable that when engineers think they have things nicely under control, a new area in technology emerges to challenge old alliances. The upstart in this case is *fluid power*. Fluid power is the technology of the transmission, control, and storage of *energy* by means of a pressurized fluid in a closed system. In the two earlier cases, the fluid itself, or its effect on an object immersed in it as an environment, was the primary concern. In the case of fluid power, the *transmission of energy* is the major concern, and the fluid enters the picture only as the means of accomplishing it. Fluid properties and behavior are of importance only as they affect transmission phenomena.

Material in current fluid mechanics texts follows this general pattern of evolution. Either the texts use basically the civil engineering approach to hydraulics with occasional reference to hydraulic machinery, or they follow the newer and more generalized approach with heavy emphasis on compressible fluid flow.

Although some of these texts are excellent, they still leave something to be desired from the point of view of fluid power. In some instances the books are so heavily oriented toward the civil engineering discipline that the mechanical engineering student must sift out the material he is concerned with. In other cases, the authors have tried to be all things to all people! They have included so much material that frequently on such a level it is impossible to do more than survey the subject matter in a one-semester introductory course in fluid mechanics. Perhaps these authors had intended their books to be used for several courses—a series in fluid mechanics. However, most schools require only *one* basic course in fluid mechanics. Thus an author cannot expect the student to receive more than a one-semester exposure to his work. For these reasons I have decided to produce this text on fluid mechanics from a different point of view from that of my colleagues.

My premise is that fluid mechanics has become so large an area of technology that it cannot be covered in its entirety in a one-semester introductory course. A large number of engineers and engineering technicians will go into industry to work in the productive segment of our economy. For the most part, these people will have been disciplined in the mechanical and electrical engineering principles. Of this group, a large majority will become involved, in one way or another, in the rapidly emerging fluid power technology. A lesser number will go into the process industries, and they will need a knowledge of fluid mechanics as related to transfer of fluids within these industries.

If this book is to be something more than just another title on the already large list of good texts on fluid mechanics, it must meet the specific need implied above. That is, it must be oriented toward the needs of the manufacturing and process industries and must place particular emphasis on fluid mechanics related to *fluid power*. It must contain enough fundamental fluid mechanics to fulfill the requirements of the individuals who may later work with fluid machinery or in the process industries.

A further consideration is that the text is designed to give a presentation of fluid mechanics within the usual one-semester, three-credit-hour course. Students requiring more work in fluid mechanics must seek it in elective courses or during graduate work. Such is the situation as it actually exists.

Since this is, supposedly, the first fluid power mechanics text, another consideration in its structure is that it must fill a wide spectrum of situations of educational application. The material presented has been classroom-proven to be useful both in ECPD-type of engineering technology curricula and as the introductory course in the four-year baccalaureate engineering curricula. Evidence of the veracity of this claim is the fact that my students have even used my lecture notes—from which this book is drawn—to pass the fluid mechanics work in the P. E. examination. I have found that in introducing a new field to students, the *concepts*, not the level of execution, cause the greatest difficulty. Of course, variation in the level of presentation can be made at the discretion of the teacher.

One last consideration—and to me the most important one—is to make the introduction to fluid mechanics as palatable for the student as possible. I do not see the merit in making something unnecessarily difficult. In this respect I take exception to the approach of generalizing everything on the presumption that the student will then be able to make the necessary transition to the specific case. It has been my observation that the beginning student does not retain enough of the abstract generalization to be able to attack the specific. He cannot relate the two because he does not have enough experience to formulate a frame of reference from which to work. Thus I prefer to start with a specific physical system which the student can visualize and understand, and, where necessary, to work back to the generalization from that vantage point. The role of the educator, after all, is to educate.

Acknowledgments

I would like to take this opportunity to express appreciation for the efforts of others who have helped bring this project to fruition. Thanks to Mr. Robert Creamer, of Temple University Technical Institute, for his valuable critiques during the preparation of the manuscript; to the editors of *Product Engineering* for permission to use material published in their magazine; to ASME for permission to use technical data originally presented in ASME publications; to Mr. Gary Thomas, faculty member at Milwaukee School of Engineering, and Mr. Torstan Trollsas, Ingeström Fellow at MSOE, for their efforts in checking out the homework problems presented at the end of each chapter; to Mrs. Josephine Gartman for preparation of the final manuscript; finally, and most important, to my students who contrived, over the years, to teach me enough so that this work might be undertaken.

R. W. H.

Milwaukee, Wisconsin
May, 1966

Contents

CHAPTER 1

An introduction to fluid mechanics

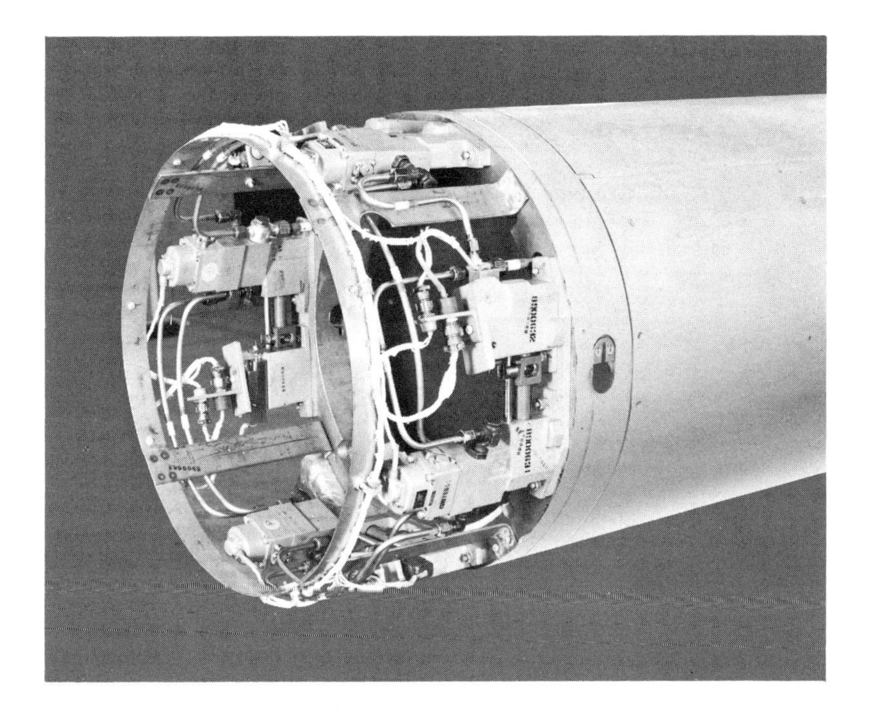

Fig. 1–1
The hydraulic servo-system used in a typical missile application.
Most rockets and missiles—including the Mercury, Gemini, and Apollo series shots—make wide use of fluid power. The supporting equipment on the ground is almost in the same position as construction equipment.
It is hard to visualize how some of the jobs could be done by any means other than fluid power.

What Is Fluid Mechanics?

We are about to indulge in an educational experience. The value of that experience to the reader will depend in great measure on his attitude toward the subject to be explored—*fluid mechanics.*

> **Fluid mechanics** is the study of the physical behavior of fluids and fluid systems and the laws describing this behavior.

The topic is not unduly difficult. In general, the behavior of fluids follows laws similar to those which describe the behavior of solid bodies and which are familiar to you from other courses, such as statics or dynamics. The major difference lies in the interpretation which must be placed on the laws due to the nature of the substance with which we are dealing. One of the most exciting areas of application of fluid mechanics is in the aerospace technology. Figure 1–1 illustrates a typical missile application.

What Is A Fluid?

Let us begin by defining a fluid:

> A **Fluid** is a substance which has definite mass and volume but has no definite shape; a fluid cannot sustain a shear stress under equilibrium conditions.

To be more precise there is one minor addition which must be made to the above definition to cover all possible cases. We should say that a fluid has definite mass and volume at constant temperature and pressure. More about this later.

There are two basic classes of fluids—*liquids* and *gases.* As students of engineering technology, we must develop discipline in our thinking. Contrary to general usage, we must include both classes when we speak of *fluids.*

> **Liquids** are fluids which have definite volumes independent of the shape of the container.

Under conditions of constant temperature and pressure, liquids will assume the shape of the container and fill a part of it which is equal in volume to the quantity of liquid. Liquids are generally thought of as being incompressible; that is, their volume doesn't change with change in pressure. Within the changes in temperature normally encountered in fluid power engineering, the volume can be thought of as being constant with regard to temperature. (Variations from these "ideal" properties will be discussed later.) Liquids, when exposed to atmospheric pressure, for example, have a free surface, such as the surface of a lake or a pail of water.

Gases are fluids which are compressible; gases will vary in volume to fill the vessel containing them.

Unlike liquids, which have a definite volume for a given mass of the liquid, the volume of a given mass of gas will change to fill the container. This behavior is described by the gas laws, which we will discuss in Chapter 17. Gases cannot have the free surface which liquids display. We will see later that there are certain conditions under which gases can also be considered incompressible, but these relate to fluid systems and not to the substance itself.

Without sacrificing accuracy, we can establish some useful relationships and develop our thinking about fluid mechanics by considering what we will call an "ideal liquid." An ideal liquid is completely incompressible; its volume does not change with temperature; and it exhibits no losses due to friction, etc., when it flows in a system. Now we know perfectly well that real liquids do not have this perfect behavior. We can, however, develop many useful relationships on this basis and then insert the above factors as corrections. In the meantime we have not cluttered up our minds with these fine points while we are developing broad, basic principles. Until further notice, then, we will be dealing with our "ideal liquid."

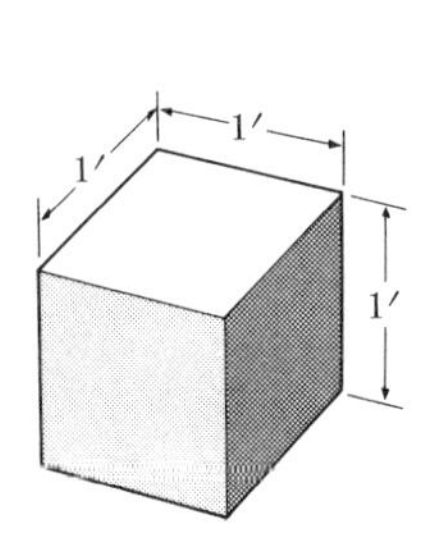

Fig. 1–2

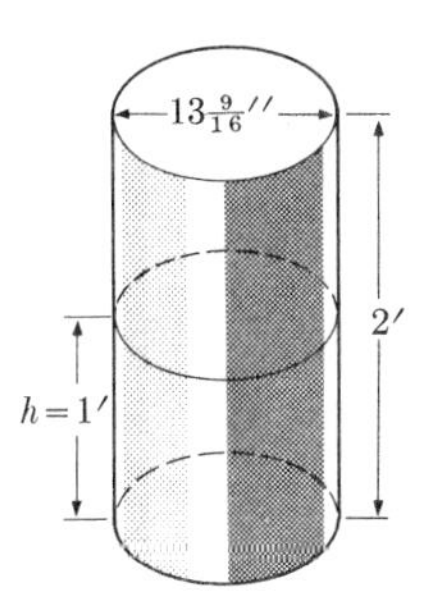

Fig. 1–3

Some Characteristics of Fluids

We have spoken about liquids having definite volume but having no shape. To visualize this, we note that the solid of Fig. 1–2 has a volume of 1 cubic foot (ft^3) because it measures 1 ft on each edge. If we have a container, or vessel, which measures 1 ft on each inside edge and if we fill it with a liquid, there will be 1 ft^3 of liquid in the container. If we have a cylindrical tank (Fig. 1–3) of about $13\frac{9}{16}$ in. in diameter (an area of about 1 ft^2) and 2 ft high and if we pour the liquid from the square vessel to the cylindrical tank, how high will the liquid rise? Well, we started out with 1 ft^3 of liquid.

$$\text{Volume} = \text{Area} \times \text{Height};$$

so the liquid will rise to a height of 1 $ft^3 \div 1\ ft^2 = 1$ ft. If the area of the base of the cylindrical tank had been $\frac{1}{2}\ ft^2$, the height of the liquid would have been 2 ft; the volume, you see, remains constant.

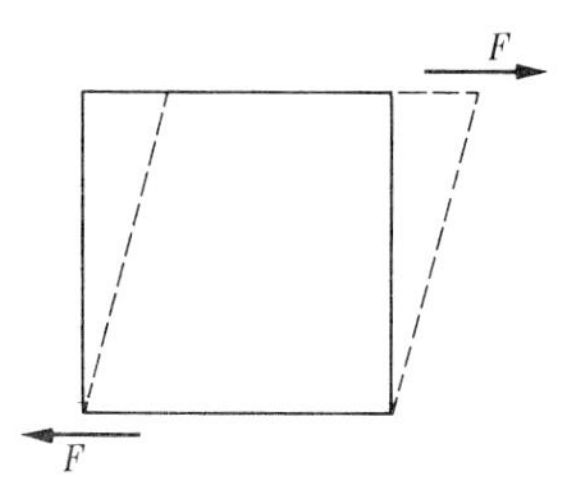

Fig. 1–4

Figure 1–4 illustrates how our solid having a volume of 1 ft^3 would react to tangential forces applied in opposite directions at the top and bottom surfaces. Such forces would induce shear stresses in the solid, such that

$$S_s = \frac{F}{l \times w} = \frac{F}{\text{area}}.$$

If we could imagine an isolated "block" of fluid similar to the solid shown in Fig. 1–4, we would find that the fluid could not sustain these two forces in equilibrium. If it could, it would be possible to pour the cubic foot of liquid out onto the table top and it would retain its shape. The fact that it would quickly spread out over the table top and seek a common level is evidence of the fact that it cannot sustain a shear stress. Note that even while it is running all over the table top, the total volume of liquid involved is still just 1 ft^3. All that has happened is that the shape of the container has changed.

Let us return to our solid block of Fig. 1–4. If it were an iron block, it would weigh about 490 lb. Since it is a 1-ft^3 block, its weight density would be 490 lb/ft^3.

If we take water as our liquid having a volume of 1 ft^3, its weight density would be 62.4 lb/ft^3. It is general practice to call this quantity the specific weight; that is,

$$\gamma = \text{Specific weight} = \text{Weight density} = \text{Weight/unit volume}.$$

The Greek letter *gamma*, γ, is used as the symbol for specific weight. For water: $\gamma = 62.4\ \text{lb/ft}^3 = 0.0361\ \text{lb/in}^3$.*

You are undoubtedly familiar with the relationship

$$M = W/g.$$

That is,

$$\text{mass} = \text{weight/acceleration due to gravity.}$$

If we use the specific weight, γ, in this equation, M is the density:

$$\text{Density} = \text{Mass/unit volume} = \gamma/g = \rho.$$

The Greek letter *rho*, ρ, is used as the symbol for density, and g is the symbol for the acceleration due to gravity, which is taken to be 32.2 feet per second per second (ft/sec^2). To distinguish between mass density and weight density, we shall hereafter refer to the former as *density* and the latter as *specific weight.*

Let us recall Newton's second law, which can be written in the form $F = Ma$. We can rewrite this as $W = Mg$, where W is the weight in pounds (force due to the acceleration of gravity), M is the mass expressed in slugs, and g is the acceleration due to gravity (32.2 ft/sec^2). All units are expressed in the English system. If we further reduce this to unit volume base, the expression becomes $\gamma = \rho g$.

What Is Pressure?

Suppose, now, that we take the cubic-foot block of iron weighing 490 lb and support it on a steel bar with a cross sectional area of 1 in^2, as shown in Fig. 1–5. What will be the reaction on the bar? It is obvious that it will be the weight of the block (490 lb). How is the reaction distributed over the steel bar? It is the weight divided by the cross-sectional area; that is, 490 lb/in^2 = 490 psi. If the block were supported on 4 rods each of cross section 1 in^2, the distributed reaction would be $490/4 = 122.5$ psi. Let us suppose that the block is just resting on a surface, such as the floor of a room. The weight (reaction) of the block is then distributed over the whole surface which is in contact with it. Since this is a 1-ft^3 block, the surface of any one face is 1 ft^2. Thus the distributed reaction (pressure) of the weight of the block over this surface is 490 lb/144 in^2 = 3.4 lb/in^2 = 3.4 psi.

---

* See Appendix B for tables of properties for other fluids.

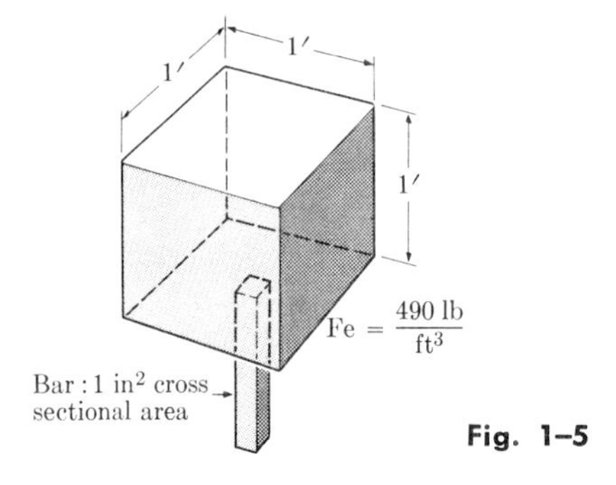

Fig. 1–5

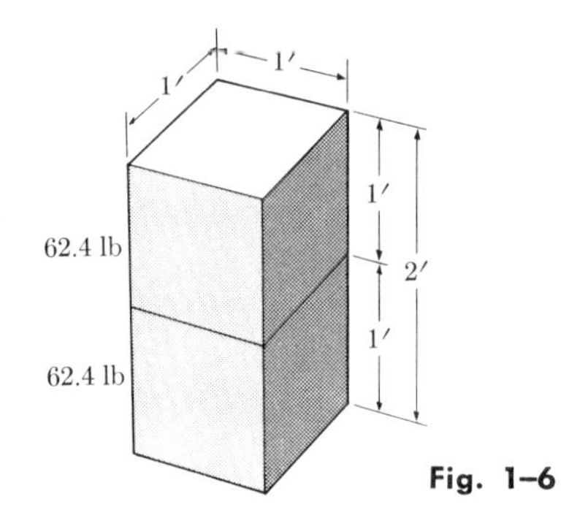

Fig. 1–6

If we now make the transition from the iron block to the 1 ft^3 of water, we find that it would weigh 62.4 lb. If we could support this 1-ft^3 volume of water on a 1-in^2 rod as we did with the iron block, the reaction would be 62.4 lb/in^2 = 62.4 psi. When we consider the distributed reaction (pressure) it becomes 62.4 lb/144 in^2 = 0.433 psi. Each square inch of surface of the bottom of the 1-ft^3 "block" of water supports 0.433 lb of distributed reaction. Note that this also represents the weight of a column of water 1 ft high above the area of 1 in^2. Thus we can see that the distributed reaction (pressure) is not dependent on the area but only on the height of the column of liquid above the measuring point. For example, consider Fig. 1–6. Here we have stacked another cubic foot of water on top of the first one. We have two 62.4-lb "blocks" being supported by the surface which previously carried only one. Thus the total reaction is $2 \times 62.4 = 124.8$ lb, and the distributed reaction, or pressure, is 124.8 lb/144 = 0.866 psi.

This gives us a hint as to how we might determine the pressure under any set of conditions. The total weight, W, of water is the product of the area of the surface, A, the height of the column, h, and the specific weight, γ. Thus,

$$W = \gamma \times A \times h.$$

Note that a *consistent set of dimensions* must be used. That is, if W is to be in pounds, then A must be in square feet, h in feet, and γ in pounds per cubic foot:

$$W(\text{lb}) = [\gamma(\text{lb/ft}^3)] \times [A(\text{ft}^2)] \times [h(\text{ft})].$$

Note that the units of feet cancel out leaving only pounds. An important point is that the verbal expression *pounds per cubic* foot really means total pounds divided by cubic feet of volume, or pounds/volume. This is very important in checking out the units, as we did in the example above. Keeping units consistent, particularily in the English system, is one of the more difficult tasks in solving problems in fluid mechanics.

Returning to Fig. 1–6, we now find that we have determined the total weight of water to be $W = \gamma A h$ lb. The pressure at the bottom surface, the distributed reaction, will be the total weight divided by the surface area, or $P = W/A = \gamma A h/A = \gamma h$ lb/unit area. If we are working in units of pounds and feet, the unit of P will be pounds per square foot, or psf. If we are working in inches and pounds, the unit of P will be pounds per square inch, or psi. Again take note of the fact that psi really means lb/in^2 and, when you check dimensionals, psi must be written in this form in the equation.

Suppose that the fluid we are considering is not water. This brings into play a factor termed specific gravity, S_g.*

Specific gravity is the dimensionless ratio of the weight of a given volume of a substance to the weight of an equal volume of water.

It is, as you can see, the ratio of their specific weights, or densities, as well. For example, we know that the specific weight of water is 62.4 lb/ft^3, and the specific weight of gasoline is 42 lb/ft^3. From the above definition, we can find the specific gravity of gasoline:

$$S_g = \frac{42 \text{ lb/ft}^3}{62.4 \text{ lb/ft}^3} = 0.675.$$

For the iron block:

$$S_g = \frac{490 \text{ lb/ft}^3}{62.4 \text{ lb/ft}^3} = 7.86.$$

Specific gravity need not be less than one. If you know the weight of the liquid (other than water) you do not need to use S_g in the calculation of pressure, as in Fig. 1–6. If you don't know it, then the relationship $\gamma_l = \gamma_w \times S_g$ can be used.

So far we have been considering an isolated "block" of water. It is practical to do so only when we include the container which holds the water. On the other hand, it is possible to mentally "isolate" our block of water in a large mass of liquid, such as a reservoir or lake. Figure 1–7 illustrates this concept.

If our block is located such that its top surface is at the level with the surface of the lake, A, it can be considered the same as the isolated block of Fig. 1–4. The block of water is supported now by the water beneath it.

---

* See Appendix B for a table of specific gravities of commonly encountered fluids.

This water below must push upward with a force equal to the weight of our isolated block pushing down. Thus, if we imagine a thin surface separating the bottom of our block from the water beneath it, the pressure (distributed reaction) pushing down on this surface would be the same as the pressure pushing up against it.

In defining a fluid earlier, we said that it was a substance which could not sustain a shear stress. What, then, keeps our "block" from "spreading out" along this imaginary surface in much the same way that it would if it were poured out on a solid surface? After all, we have established that there is a pressure distribution across the "surface." The fact is that the water adjacent to the sides of our block pushes on the sides in much the same way as the water below pushes up against it. Thus at any point on any side of our imaginary block there is a pressure pushing out and a like one pushing in—action and reaction. This leads us to a very important principle, one which is basic to the operation of fluid power systems.

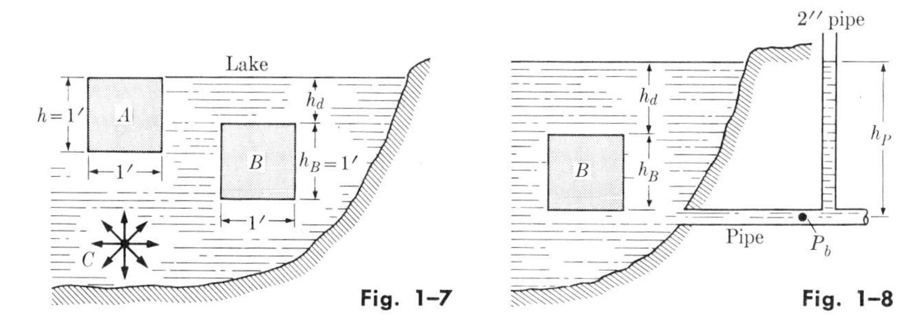

Fig. 1–7

Fig. 1–8

Pascal's Law

Pascal's law tells us that the pressure at any point in a fluid, say point C in Fig. 1–7, is equal in all directions.

The pressure in a static fluid is the same in all directions; pressure applied to a fluid is transmitted undiminished throughout the fluid.

Let us take a look at our imaginary block at point B in Fig. 1–7. Here the surface of the block is no longer coincident with the surface of the lake. We established earlier that $P = \gamma h$. In our submerged block we must consider not only h_B, the height of our block, but also h_d, the depth of the block beneath the surface of the lake. Thus the pressure on the top surface of our block would be $P_t = \gamma h_d$. That at the bottom surface would be $P_b = \gamma(h_B + h_d)$. The difference in pressure between the top and the bottom surfaces would be the same as before: $\gamma h_B = \gamma(h_B + h_d) - \gamma h_d$. Thus we have reaffirmed the statement made earlier that the pressure at any point in a fluid is a function of its depth (and the density of the fluid). If point C

in Fig. 1–7 were at the same level as the bottom of our block, the pressure would be P_b. This can be expanded to include all the points in the lake: the pressure at any point on the *same level* is the same. Or we could say that we have constant pressure at constant depth h.

Let us suppose that we push a pipe 2 in. in diameter down into the lake so that its lower end is at the same level as the bottom of our imaginary block of water. What would be the pressure in the end of the pipe? It would be

$$P_{b(\text{block})} = P_{b(\text{pipe})} = \gamma(h_B + h_d).$$

If we had a piece of pipe of length equal to $h_B + h_d$, and if we filled it with water, would the pressure still be P_b? Of course it would. Figure 1–8 depicts our lake with a pipe leading from it at the same depth as the bottom of our block. The 2-in. pipe is connected to the horizontal pipe outside the lake. As determined before, the pressure at this level is P_b. According to Pascal's law, the pressure in the horizontal pipe leading from the lake will be P_b. How high will the water rise in the vertical 2-in. pipe? From the relationship $P_b = \gamma(h_B + h_d)$, we derive

$$h_B + h_d = \frac{P_b}{\gamma} = h_P.$$

Pressure Head

The height of water, h_P, is given a special name *pressure head.* Pressure head is the vertical height, in feet or inches, to which a given pressure will elevate a column of fluid. For every pressure level there is a corollary pressure head. In a massive body of fluid, such as a lake or the atmosphere, depth in the fluid and pressure head are synonymous. For a piped system operating at any pressure, pressure head is the height to which the column of liquid could be raised. Note the difference in units between pressure (psi) and pressure head (in.).

IMPORTANT TERMS

A **fluid** is a substance which cannot sustain a shear stress under equilibrium conditions.

Liquids and **gases** are classes of fluids.

Pressure is force per unit area:

$$P = \frac{F}{A}\ \text{lb/in}^2\ (\text{or lb/ft}^2).$$

Specific weight is the weight per unit volume:

$$\gamma = \frac{W}{V} \text{ lb/in}^3 \text{ (or lb/ft}^3\text{)}.$$

Density is the mass per unit volume:

$$\rho = \frac{M}{V} = \frac{W/g}{V} \text{ slugs/in}^3 \text{ (or slugs/ft}^3\text{)}.$$

Specific gravity is the dimensionless ratio of the weight of a substance to the weight of the same volume of water.

Pascal's law: pressure is transmitted undiminished throughout a fluid.

Pressure head is the vertical height to which a given pressure will elevate a column of fluid. It is expressed in units of feet or inches.

PROBLEMS

1–1 What is the difference between a *fluid* and a *solid?*

1–2 What differentiates *liquids* from gases?

1–3 What is the difference between *specific weight* and *specific gravity?*

1–4 If an oil has a specific weight of 0.0300 lb/in^3, what is its specific gravity?

1–5 What is the difference between *specific weight* and *density?*

1–6 What is the density of the oil of Problem 1–4?

1–7 If we have a fluid of $S_g = 1.15$, what is its density?

1–8 What height would a column of mineral oil have to be to exert the same pressure at its base as an 18-in. column of mercury?

1–9 In the system shown in Fig. 1–9, assume that the iron block is frictionless in the standpipe and that *no* leakage occurs. Calculate the pressures at the different levels.

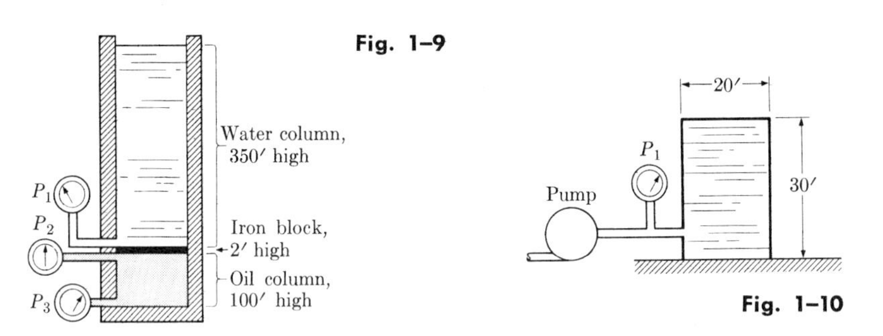

Fig. 1–9

Fig. 1–10

1–10 Using the principles developed in this chapter, prove the following expressions: (a) $P = 0.433 \times S_g \times h$, where P is pressure (psi), S_g is the specific gravity of a liquid, and h is the head (ft). (b) $h = (2.32 \times P)/S_g$, with the same units.

1–11 A skin diver descends 20 ft into the ocean. The specific gravity, S_g, of salt water is 1.03. What is the pressure, in psi, to which he is subjected?

1–12 A tank is shaped like the frustum of a cone. It is 5 ft in diameter at the bottom, 2 ft in diameter at the top, and 10 ft deep. If it contains glycerine ($S_g = 1.260$) to a depth of 8 ft, what is the pressure, in psi, on the bottom?

1–13 Pressure in a given liquid varies according to the expression $P = \gamma h$. Draw a plot of the variation in pressure from the surface of a lake to the bottom.

1–14 A 100-ft pipe rests on the side of a hill at a 30° angle with the horizontal. If it is filled with benzene, $S_g = 0.899$, what is the pressure at the base of the pipe?

1–15 Pressure in a 12-in. water main is 150 psi. The main is buried 15 ft below street level. How high above street level can water be delivered?

1–16 Describe how the principle of *pressure head* might be used to measure pressure in a pipe when a tube is inserted into the wall of the pipe.

1–17 A cylindrical tank 100 ft in diameter and 50 ft deep stands on top of a hill 75 ft high. The tank filled with water supplies water mains buried 15 ft below the base of the hill. What will the pressure be at water valves on the second floor of a manufacturing plant which is 20 ft above the base of the hill?

1–18 A drain pipe is located in the side of a tank 15 ft from the top. Its inside diameter is 6 in. The fluid in the tank is water. A workman must hold a flat cover flange over the end of the pipe while it is bolted in place. How hard will he have to push to hold it?

1–19 In Problem 1–17, what pressure must pumps develop to pump water into the storage tank from a pit in the pump house which is 20 ft below the base of the hill?

1–20 A submarine is cruising at a depth of 320 ft off the Florida coast. The captain wishes to pump out the ballast tanks. What pressure must the pumps develop?

1–21 A typical storage tank is shown in Fig. 1–10. The pump is delivering oil, $S_g = 0.81$, into the tank. A careless workman allows the tank to fill completely and lets the pressure, P_1, at the base of the tank build up to 12 psi. What force is exerted on the top of the tank?

CHAPTER 2

Pressure, head, force

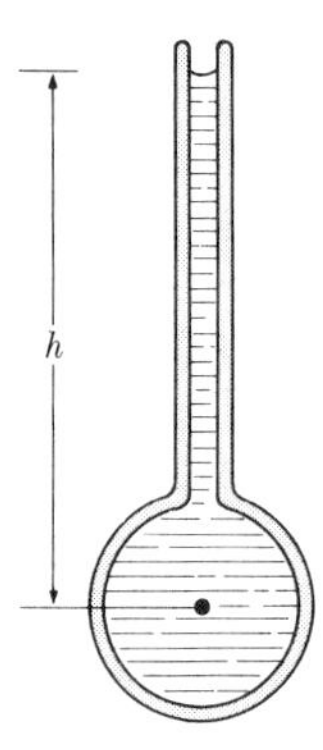

Fig. 2–1

Pressure Head and the Piezometer Tube

The device illustrated in Fig. 2–1 is called a *piezometer tube* and is used to indicate pressure in a conductor. The liquid will rise in the tube to a height equal to the pressure head; $h = P/\gamma$, where h is in ft (or in., etc.), P is in psf (or psi), and γ is in lb/ft^3 (or lb/in^3). Check this relationship dimensionally:

$$h(\text{ft}) = \frac{\text{lb}}{\text{ft}^2} \div \frac{\text{lb}}{\text{ft}^3} = \frac{\text{lb} \cdot \text{ft}^3}{\text{lb} \cdot \text{ft}^2} = \frac{\text{ft}^3}{\text{ft}^2} = \text{ft}.$$

This technique of dimensional checking is an invaluable tool to learn to use.

The piezometer tube can be used to measure relatively low, positive pressures. It also serves as an excellent illustration of the equivalency of pressure and pressure head.

In Problem 1–1, we derived a rule of thumb relationship for the calculation of head or pressure and used the more familiar units: psi for pressure, and ft for head. Thus

$$P = 0.433 S_g h \qquad \text{and conversely} \qquad h = \frac{2.32P}{S_g}.$$

Absolute and Gage Pressure

We stated above that a piezometer tube could best be used to indicate positive pressure. One might well ask what other kind of pressure there is and from a purely algebraic sense would arrive at the conclusion, negative pressure. Since we defined pressure as *force per unit area*, a force of something less than nothing may be a little difficult to envision. Figure 2–2 depicts the method by which we may have either positive or negative pressures. The earth is enclosed in an environment of air, the atmosphere. We cannot do anything on the surface of the earth without it being done relative to this environment. It is somewhat analogous to our being at point C in Fig. 1–7; we are surrounded by a "lake" of air. The atmosphere has a depth just as assuredly as does the lake of Fig. 1–7. This depth is the same as the pressure head. In other words, the atmosphere has weight, and the distributed reaction of this weight on the earth's surface is what we call *atmospheric pressure.* It is well known that this pressure is about 14.7 psi at sea level.

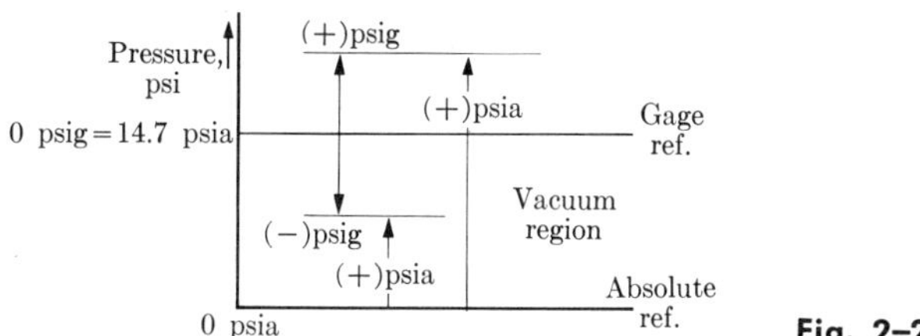

Fig. 2–2

It is impossible to make a pressure measurement on the earth's surface unless it is made relative to atmospheric pressure. Therefore a reference can be established at the atmospheric pressure level, as indicated in Fig. 2–2. If the pressure we wish to measure is at the same level, there will be zero pressure relative to atmospheric. Pressure gages, piezometers, and other pressure-measuring devices indicate pressures called *gage pressures.* Atmospheric pressure is 0 psi gage, often abbreviated psig. Gage pressures are positive if they are greater in magnitude than atmospheric, and are negative if less than atmospheric and measured down from the atmospheric reference. Negative pressures are also called *vacuums.*

If we had a negative pressure of −14.7 psig, it is obvious that we would have no pressure at all. We know that atmospheric pressure is 14.7 psi at sea level. Then 14.7 − 14.7 = 0. This condition of no pressure at all is called *absolute zero.* It is analogous to absolute zero temperature in the measurement of temperature. Thus we have a second reference from which pressure can be measured, absolute zero or 0 psia. We can see now that atmospheric pressure is actually 14.7 psia = 0 psig. Vacuums can be expressed in either absolute or gage terms. For example, in Fig. 2–2, we see that −5 psig = 9.7 psia. It is general practice to mean "gage pressure" when using the abbreviation psi. It is necessary to use the term "psia" when absolute pressure is being expressed.

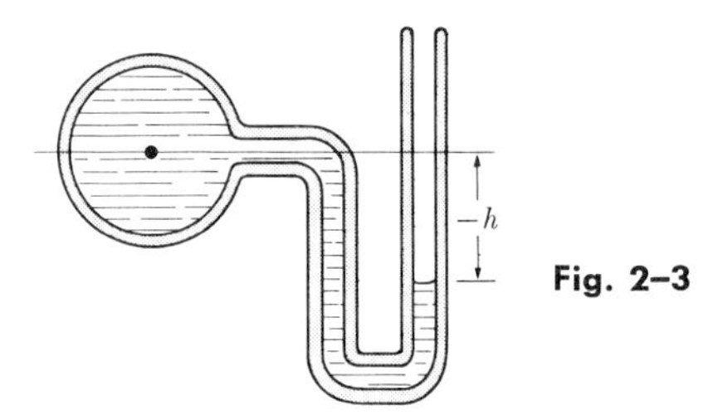

Fig. 2–3

Pressure Head and the Manometer

It is now apparent why the piezometer can best measure positive pressures. If a negative pressure existed in the pipe, atmospheric pressure would push down into the pipe. In many industrial applications such aeration is intolerable.

The piezometer with a U-tube, as shown in Fig. 2–3, overcomes part of this problem. The reference is at the opening into the pipe: positive pressure would result in a head measured above the reference; vacuums would result in a head below.

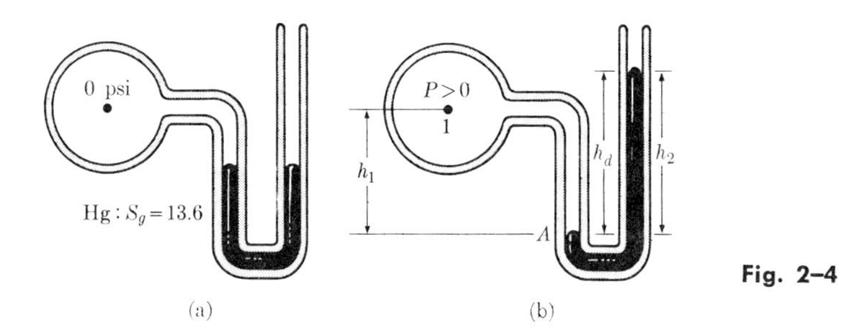

Fig. 2–4

A *manometer* is a pressure-measuring device which has a gage liquid different from that in the pipe. Mercury is most frequently used; water finds application when low pressures are being measured. Figure 2–4 shows a typical manometer configuration. When the pressure in the pipe is 0 psi, the mercury in the U-tube seeks its own level and the two legs are of equal length, as shown in Fig. 2–4(a). When pressure, assuming it is positive, is applied in the pipe, the force on the left-hand leg pushes the mercury up in the right-hand leg, as shown in Fig. 2–4(b). Note that the reference is again taken at the level where the manometer tube enters the pipe. For the system to be in equilibrium, the weight of the column of liquid in the right side must equal the weight of the column of liquid in the left side.

In other words, the pressure at any level in the right-hand leg must be the same as the pressure at the same level in the left-hand leg. A convenient point is the interface between the two fluids, point A in Fig. 2–4(b). Then, considering the left leg of the manometer, we have

$$P_{a_R} = P_1 + \gamma_1 h_1,$$

and in the right leg,

$$P_{a_1} = \gamma_2 h_2.$$

If equilibrium exists, that is, if the manometer isn't moving, we have $P_{a_R} = P_{a_L}$, and hence

$$P_1 + \gamma_1 h_1 = \gamma_2 h_2, \qquad \text{or} \qquad P_1 = \gamma_2 h_2 - \gamma_1 h_1.$$

If the specific gravities of the fluids are known, the above equation can be rewritten as follows:

$$P_1 = \gamma_w (S_{g_2} h_2 - S_{g_1} h_1),$$

where γ_w is the specific weight of water.

Manometers are used in accurate determination of relatively low pressures. For example, even a manometer 10 ft high would be capable of measuring only

$$P = \gamma h = S_g \gamma_w h = 13.6 \times 62.4 \text{ lb/ft}^3 \times 10 \text{ ft}$$

$$= 8486.4 \text{ psf} = 58.9 \text{ psi}.$$

Note that 1 in. of mercury is approximately equivalent to 0.49 psi; that is,

$$P = \gamma h = (13.6 \times 0.0361) \times 1 \text{ in.} = 0.49 \text{ psi}.$$

From this simple relationship we can also determine the height of the column of mercury which can be supported by atmospheric pressure: $h = P/\gamma = 14.7 \text{ psia}/0.49 \simeq 30$ in. Hg. It is a commonly recognized fact that atmospheric pressure will support about a 30-in. column of mercury. Note that this is in reality the pressure head expressed in terms of inches of mercury.

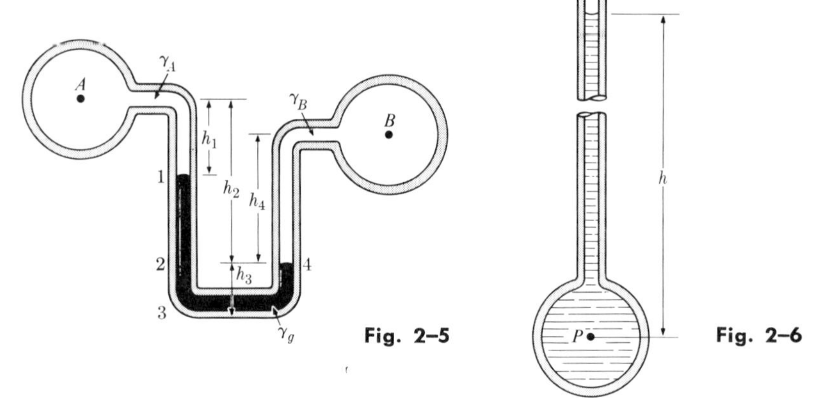

Fig. 2–5

Fig. 2–6

The Differential Manometer

The mercury manometer of Fig. 2–4 is used to measure absolute pressures relative to the atmospheric environment. The differential manometer, illustrated in Fig. 2–5, is used to measure the pressure (head) difference between two reference points in fluid media; most frequently between two pipes as shown. In contrast to the simple manometer of Fig. 2–4, the differential manometer *cannot* indicate absolute pressure; it can register only the difference between the two pressures. Solving for the pressure difference as indicated by the differential manometer is an excellent exercise for equating pressure and head. It is important to recognize that when the manometer has stopped fluctuating the various fluids within the system are in equilibrium. We can set up a chart of pressures (or equivalent heads) on the right side of the manometer and on the left side.

LEFT SIDE		RIGHT SIDE	
point	*pressure*	*point*	*pressure*
A	P_A	B	P_B
1	$P_A + \gamma_A h_1$		
2	$P_A + \gamma_A h_1 + \gamma_g(h_2 - h_1)$	4	$P_B + \gamma_B h_4$
3	$P_A + \gamma_A h_1 + \gamma_g(h_2 + h_3 - h_1)$	3	$P_B + \gamma_B h_4 + \gamma_g h_3$

We can see that the two expressions for the pressure at point 3 must be equal. Thus

$$P_A + \gamma_A h_1 + \gamma_g(h_2 + h_3 - h_1) = P_B + \gamma_B h_4 + \gamma_g h_3.$$

The expression $\gamma_g h_3$ can be eliminated from the equation, leaving

$$P_A + \gamma_A h_1 + \gamma_g(h_2 - h_1) = P_B + \gamma_B h_4,$$

from which

$$P_A - P_B = \gamma_B h_4 - \gamma_A h_1 - \gamma_g(h_2 - h_1)$$

but $h_2 - h_1$ = difference in Hg head = h_d. Then

$$P_A - P_B = \gamma_B h_4 - \gamma_A h_1 - \gamma_g h_d$$

is the final expression for the pressure difference between the two pipes.

In fluid power systems used in industry, operating pressures in the range of 1000 to 5000 psi are quite common. If we tapped a piezometer tube into a pipe carrying water at 1000 psi, how high would the water rise in the tube shown in Fig. 2–6? We know that

$$P = \gamma h \qquad \text{or} \qquad h = P/\gamma;$$

then

$$h = \frac{1000 \text{ lb/in}^2}{0.0361 \text{ lb/in}^3} = \frac{27{,}700 \text{ in}}{12} = 2320 \text{ ft.}$$

Thus we see that the pressure head for 1000 psi pressure is 2320 ft of water. Let's look at the dimensional check of the above solutions:

$$h = \frac{\text{lb}}{\text{in}^2} \div \frac{\text{lb}}{\text{in}^3} = \frac{\text{lb} \cdot \text{in}^3}{\text{lb} \cdot \text{in}^2} = \text{in.}$$

If we look at it another way, we find that the distributed reaction of a column of water 2320 ft high would be a pressure of 1000 psi.

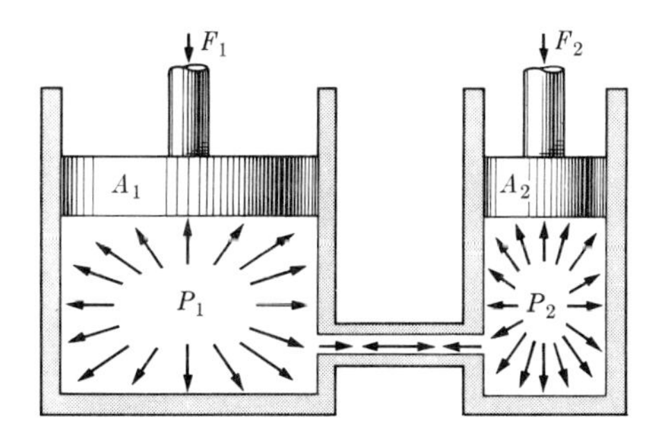

Fig. 2–7

Force Multiplication Via Pressure and Area

You will recall that Pascal's law told us that pressure is transmitted undiminished throughout a fluid. Figure 2–7 shows a device which for its function depends upon the application of Pascal's law. We have represented here two cylindrical containers and two circular pistons fitted closely to the walls of the cylinders. They are not unlike the pistons in the cylinders

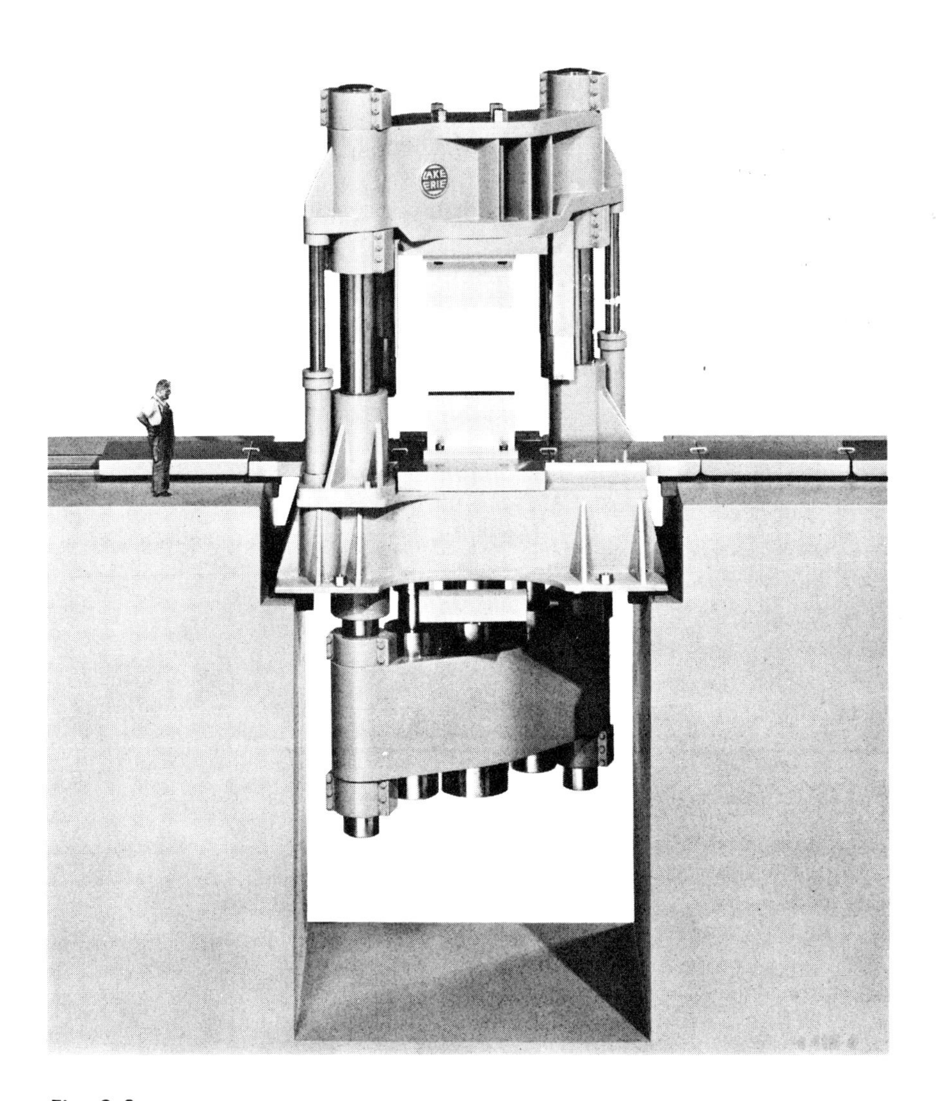

Fig. 2–8

Hydraulically powered forming press such as is used throughout industry for forming metal shapes. Columns on either side of press opening are hydraulic cylinders which generate tons of force between the blocks in the press opening. This is a practical application of Pascal's law.

of a car engine. Cylinder A is of different cross-sectional area from cylinder B. The two are connected by a pipe so that fluid could flow between them or, according to Pascal, that pressure could be transmitted from one to the other.

If we exert a force, F_1, on the first piston, the reaction of this force will be distributed over the area of the piston, A_1. Thus, $F_1 = P \times A_1$ or $P = F_1/A_1$. The pressure generated is the distributed reaction of the force over the area. Since this pressure is transmitted undiminished through the liquid, a force $F_2 = P \times A_2$ could be exerted by the second piston. This principle is the basis for all fluid power technology involving the transmission of energy by means of pressurized fluids. Typical applications with which we are familiar include hydraulic brakes on cars and hydraulic jacks, where a relatively small force applied to a small area generates a pressure. This pressure is then transmitted *undiminished* to a large area, where a huge force can be developed. Figure 2–8 illustrates how this idea is put to work in modern production industry.

We can solve the two expressions above for P and equate them as follows:

$$P = \frac{F_1}{A_1} = \frac{F_2}{A_2} \qquad \text{or} \qquad F_1A_2 = F_2A_1.$$

This shows that the force is inversely proportional to the area on which the pressure acts. Even more important, it points up that there can be no pressure generated if one of the forces or reactions is zero. Thus pressure is *not generated* as a function of the input but as a function of the output reaction or resistance. Such a function is frequently called the *load*. Thus, in the example above, if we consider F_1 to be the input, then F_2 is the load reaction. This is basic to fluid power systems.

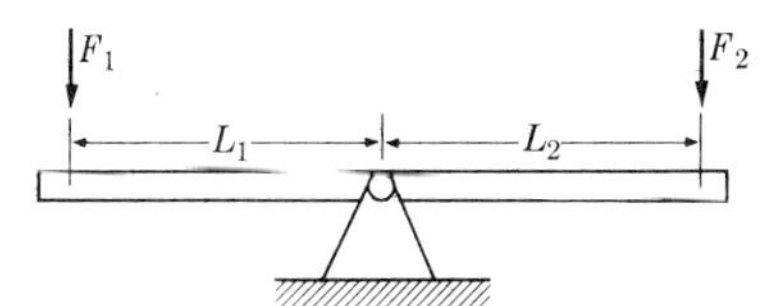

Fig. 2–9

An interesting analogy can be drawn between the fluid system of Fig. 2–7 and the mechanical lever system shown in Fig. 2–9. In the familiar lever system, we know that the product of a force and its lever (moment) arm must equal the product of the opposite force and its lever arm. That is, $F_1 \times L_1 = F_2 \times L_2$ for equilibrium to obtain. Note that the lever arms here play the same role as the areas of the pistons in the fluid system of Fig. 2–7. The lever itself functions as does the fluid; they both transmit the force but do not play an active part, *per se*, in the force balance which

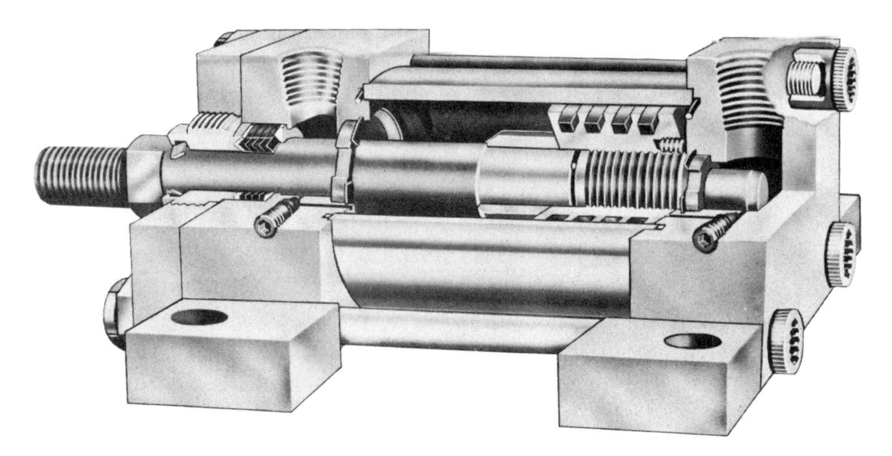

Fig. 2–10
Typical double-acting cylinder. (Courtesy of Tompkins-Johnson Co.)

ensues. One might consider the fluid as a "flexible lever" capable of transmitting the force (via pressure) independently of the geometry of the system itself.

The Fluid Power Cylinder

Cylinders, such as we have been considering, play a major roll in industrial fluid power applications. A typical design is illustrated in Fig. 2–10. Cylinders are used on machine tools, agricultural equipment, construction machinery, aircraft and missiles, and ships; hardly an area of industry exists that doesn't make use of cylinders in some way.

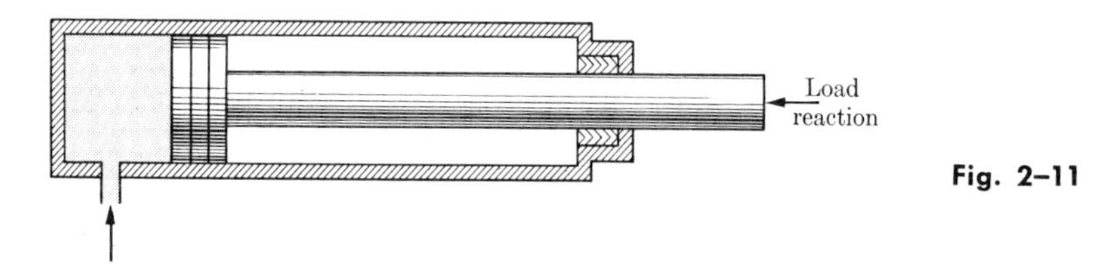

Fig. 2–11

Figure 2–11 shows a simple *single-acting* cylinder design. This means that pressurized fluid is admitted only at one end. Thus the cylinder can move in only one direction under power. If the area of the piston is A_P, and the load resistance is F_L, the equation $F_L = P \times A_P$, describes the equilibrium condition for the cylinder. If any two of the variables are known, the third can be calculated.

EXAMPLE

Suppose we wish to move a 100,000-lb load. We have a pump which can deliver a maximum pressure of 1000 psi. What diameter cylinder (bore) shall we use?

Step 1. From $F_L = P \times A_P$,

$$A_P = F_L/P = 100{,}000 \text{ lb}/1000 \text{ lb/in}^2 = 100 \text{ in}^2.$$

Step 2. $A_P = (\pi/4)D^2$, $D = \sqrt{4A_P/\pi} = \sqrt{400/\pi} \simeq 11.25$ in.

The cylinder configuration of Fig. 2–12 is called a *regenerative* cylinder. There is a pipe, or conductor, connecting the rod end of the cylinder with the blank end. Let the area of the piston be A_P. The area of the rod is A_R.

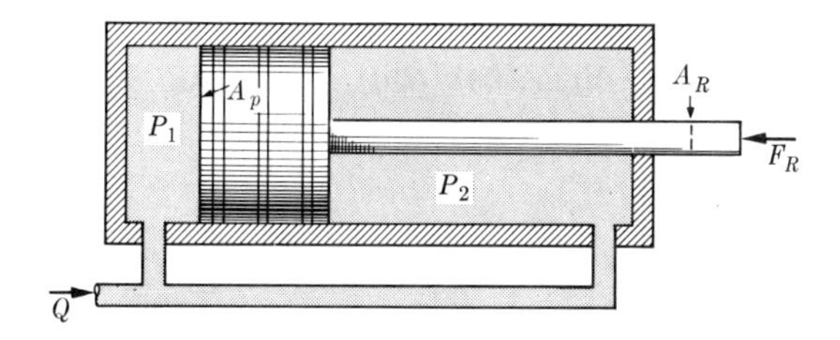

Fig. 2–12

It first appears as though the cylinder would be hydraulically locked up and could not move; however, a close analysis shows the forces acting on the piston with an inlet pressure, P, due to a load resistance, F_R. Thus

$$\Sigma F_x = 0, \qquad P_1 \times A_P - F_R - P_2 \times (A_P - A_R) = 0,$$

but $P_1 = P_2$ (Pascal's law), and the two ends of the cylinder are connected. Then

$$P_1 \times A_P - F_R - P_1 \times A_P + P_1 \times A_R = 0,$$

and

$$F_R = P_1 \times A_R.$$

Thus we see that the cylinder is actually capable of overcoming a load resistance equal to the product of the pressure and the area of the *piston rod*. Or, more accurately, the pressure developed will be equal to the load resistance divided by the area of the piston rod.

IMPORTANT TERMS

Piezometer tube is a single column or tube inserted into a pressure pipe to measure the pressure head.

Gage pressure is pressure measured relative to the earth's atmospheric pressure, the latter being considered 0 psig.

Absolute pressure is pressure measured relative to an absolute zero scale. It corresponds to no pressure whatsoever; atmospheric pressure is 14.7 psia at standard conditions.

Vacuum is the existence of a pressure level below atmospheric.

Manometer is a double-legged tube which utilizes a different gage fluid from the working fluid in a pipe to measure the pressure head.

Pascal's law states that the pressure in a fluid is transmitted undiminished in all directions.

Cylinder is a fluid-power component used for generating linear forces as a function of fluid pressure.

Regenerative cylinder is a cylinder in which both ends of the cylinder are connected in parallel. The intent is to increase the velocity of the piston rod.

PROBLEMS

2–1 A column of water is measured and found to be 1160 ft high. What is the *pressure* at the bottom of the column?

2–2 If the column of Problem 2–1 is a pipe 6 in. in diameter, what is the *force* on the base of the column?

2–3 If the base of the column of water of Problem 2–1 lies at sea level, in what units is the pressure expressed? What other units of *pressure* might you use and how would you convert from one to the other?

2–4 If the fluid of Problem 2–1 was sea water instead of fresh water, what would the *absolute pressure* be at the base of the column?

2–5 If we substituted hydraulic oil* for the water, what would the *pressure* be at the base of the column?

2–6 How high a column of mercury would we need to give the same *pressure* as observed in Problem 2–1? If the mercury column were changed from a 6 in. pipe to a 3 in. pipe, how would the *pressure* be affected at the base of the column?

---

* Refer to Table B–1, p. 219. Use "mineral oil."

2–7 Describe the effect of specific gravity on *pressure* observed at the base of a column of fluid.

2–8 A pipe carries hydraulic oil at a *pressure* of 500 psi. If a piezometer tube were to be used to measure the pressure head, how long would it have to be?

2–9 In Problems 2–1 through 2–6, what *is* the *pressure head* for each of the pressures observed or calculated?

2–10 A pipe in a fluid-power system carries a fluid pressure of —3 psi. What is the absolute pressure? What term is applied to pressures in this region?

2–11 Calculate the vacuum in inches of mercury equivalent to the pressure of Problem 2–10.

2–12 Sketch a mercury manometer which might be used to measure the vacuum of Problem 2–10.

2–13 In the manometer shown in Fig. 2–13, fluid A is benzene, and fluid B is mercury. What is pressure P_1?

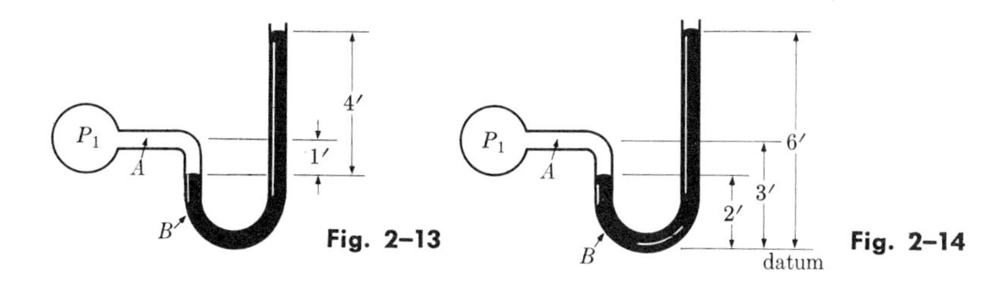

Fig. 2–13

Fig. 2–14

2–14 With the same fluids as in Problem 2–13, what will the pressure P_1 be in the manometer in Fig. 2–14?

2–15 In Fig. 2–15, fluid A is water, and fluid B is mercury. What is pressure P_1?

2–16 In the manometer of Problem 2–15, if fluid A is *air* and fluid B is water, what is the pressure P_1?

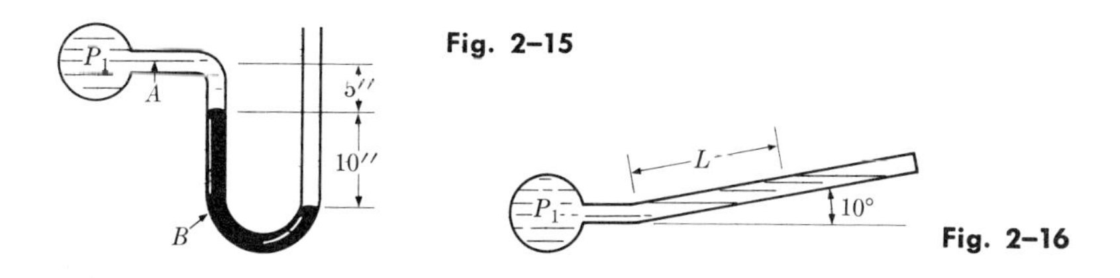

Fig. 2–15

Fig. 2–16

2–17 In the inclined piezometer in Fig. 2–16 the length L is $57\frac{1}{2}$ in. If the fluid is water, what is the pressure P_1?

2–18 By means of a sketch, show how the principle of Problem 2–17 could be applied to a manometer to measure low gas pressures. What advantage would it have over a vertical manometer?

2–19 In the differential manometer in Fig. 2–17, fluid A is water, fluid B is hydraulic oil, gage fluid is mercury, $h_1 = 10$ in., $h_2 = 40$ in., $h_3 = 4$ in., $h_4 = 32$ in. What is the pressure difference, $P_A - P_B$?

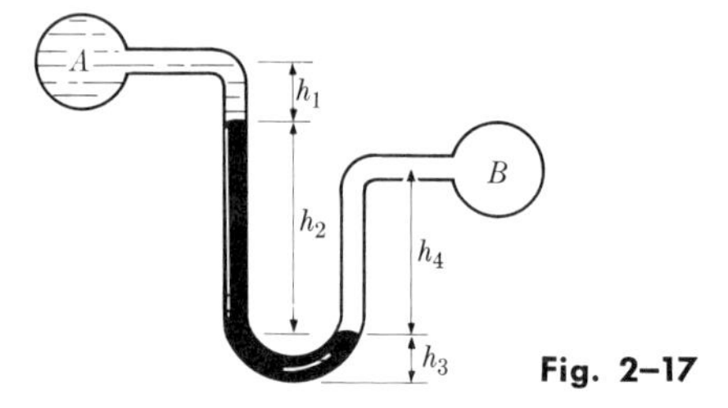

Fig. 2–17

2–20 In the differential manometer of Problem 2–19, if fluid A is water, fluid B is water, gage fluid is an immiscible hydraulic oil, $h_1 = 5$ in., $h_2 = 20$ in., $h_3 = 2$ in., $h_4 = 11$ in., then what is the pressure difference, $P_A - P_B$?

2–21 A differential manometer, such as that of Problem 2–19, has oil for fluid A, benzene for fluid B, and mercury as the gage fluid. If $h_1 = 10$ in., $h_4 = 20$ in., and $P_B - P_A = 100$ psi, what is the height h_2?

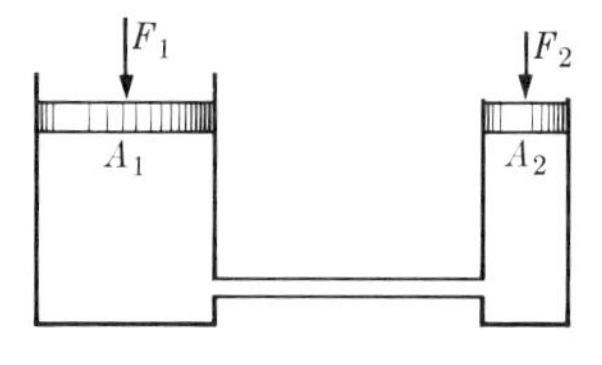

Fig. 2–18

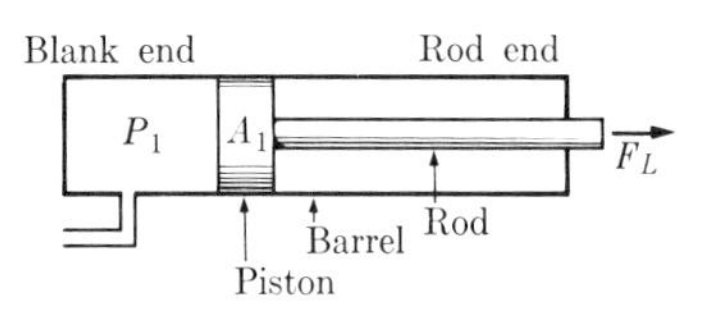

Fig. 2–19

2–22 In Fig. 2–18, which demonstrates Pascal's law, if $F_1 = 1000$ lb, $A_1 = 10$ in^2, and $A_2 = 2.5$ in^2, what is the value of F_2?

2–23 (a) In Fig. 2–18, if $F_1 = 200$ lb and $F_2 = 500$ lb, what is the ratio of the areas? (b) If $A_1 = 4$ in^2, what is A_2?

2–24 In Problem 2–23, if piston A_2 moves 2 in. in the cylinder, how far will piston A_1 move?

2–25 If we neglect losses in Problem 2–24, how much work will be done by F_1 and F_2?

2–26 Figure 2–19 shows a typical *cylinder* which is used in fluid power applications to produce a linear force by introducing hydraulic pressure behind the piston. If the pressure $P_1 = 1000$ psi, and the piston area $A_1 = 15$ in^2, what load force, F_L, can the cylinder exert?

2–27 For the cylinder in Fig. 2–19, suppose a load exerts a reaction, $-F_L$, on the piston rod of 25,000 lb. If the piston area $A_1 = 12.5$ in^2, what pressure would be necessary to just balance the load reaction?

2–28 The load reaction of 25,000 lb in Problem 2–27 is imposed on the piston rod. A maximum pressure of 4000 psi is available. What must the size of piston A_1 be?

2–29 In Fig. 2–20 two cylinders are shown connected in series. If $A_1 = 10 \text{ in}^2$, $A_2 = 4 \text{ in}^2$, $A_3 = 6 \text{ in}^2$, and $F_2 = 3000$ lb, what force, F_1, is required to bring the system into equilibrium when $P_1 = 1000$ psi?

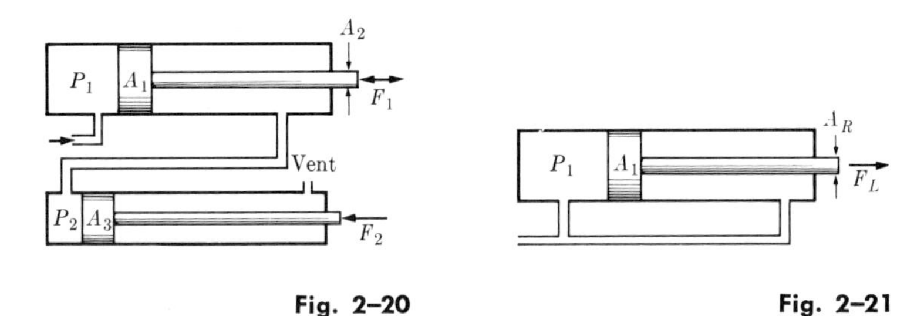

Fig. 2–20

Fig. 2–21

2–30 Using Fig. 2–20, let us assume that $F_1 = 1000$ lb, $F_2 = 2000$ lb, $A_1 = 8 \text{ in}^2$, $A_2 = 3 \text{ in}^2$, $A_3 = 5 \text{ in}^2$. What pressure, P_1, is needed to bring the system into equilibrium?

2–31 A regenerative cylinder is shown in Fig. 2–21. If $A_1 = 10 \text{ in}^2$, $A_R = 4 \text{ in}^2$, and $P_1 = 1000$ psi, what will F_L be?

2–32 Given a regenerative cylinder, such as illustrated in Fig. 2–21, in which $P_1 = 1500$ psi and $F_L = 7500$ lb, determine A_1 and A_R.

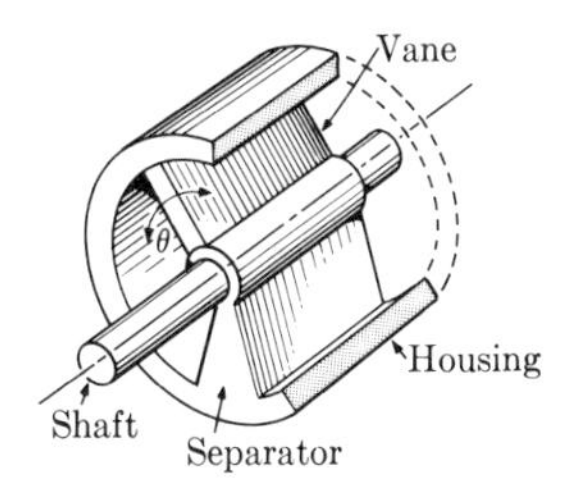

Fig. 2–22

2–33 The device in Fig. 2–22 is a fluid power component used to produce a rotational output analogous to the linear output of a cylinder. It is called a *rotary actuator*. A rectangular vane is attached to a shaft, which is supported on bearings. A *separator* divides the cylindrical housing into two sections, one on either side of the vane. When pressurized fluid is introduced into one section, an unbalanced hydraulic force is developed against the vane. Since the vane is off center relative to the shaft, a torque is produced by this hydraulic force. Let

R = outside radius of the vane,

r = inner radius of the vane,

w = width of the vane,

P = hydraulic pressure applied to the vane,

T_o = output torque.

Develop the expression for output torque as a function of the above parameters.

2–34 Given a rotary actuator, as shown in Fig. 2–22, the inside diameter of the housing = 10 in., the shaft diameter = 2 in., width of the vane = 4 in., pressure = 3000 psi. Neglecting losses, calculate the output torque of the actuator.

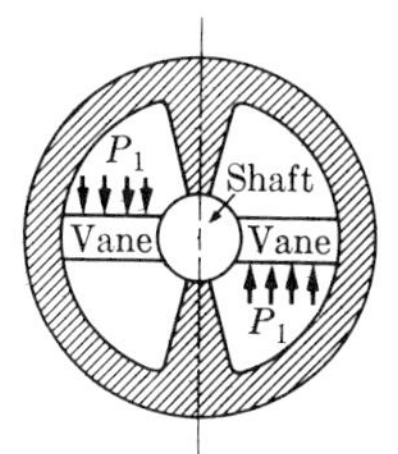

Fig. 2–23

2–35 Some rotary actuators are similar to the double-vane types shown in Fig. 2–23. Using the same parameters as in Problem 2–34, calculate the output torque for a double-vane rotary actuator.

CHAPTER 3

Buoyancy, force on submerged surfaces

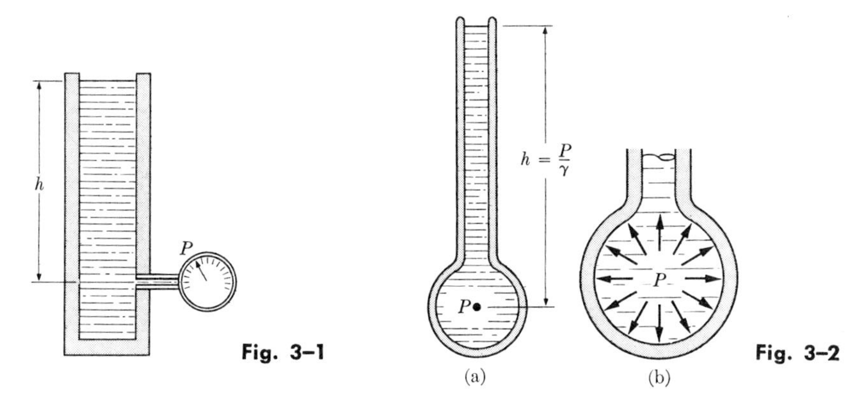

Fig. 3–1

Fig. 3–2

Hydrostatic Pressure

Hydrostatic pressure is the term frequently applied to the distributed reaction of (a) a mass of fluid on a surface or (b) an applied force transmitted through a fluid according to Pascal's law. The connotation is one of pressure due to a fluid at rest, or nearly at rest; it will be contrasted with hydrodynamic reaction in Chapter 11.

It has been established that a column of incompressible fluid of height h will generate a pressure (force per unit area) due to the weight of the fluid in the column. This is illustrated in Fig. 3–1. The pressure gage will indicate a pressure $P = \gamma h$, where P, h, and γ must be expressed in a consistent set of units. If we change the configuration of the column to that of Fig. 3–2(a), we do not alter the relationship in any way. According to Pascal's law, the pressure distribution on the wall of the pipe will be as shown in Fig. 3–2(b). Since by definition a fluid in equilibrium cannot

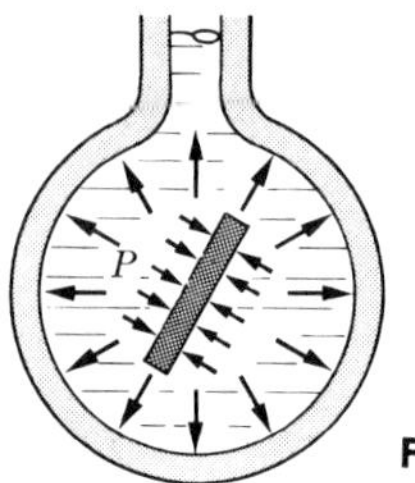

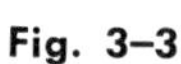

Fig. 3–3

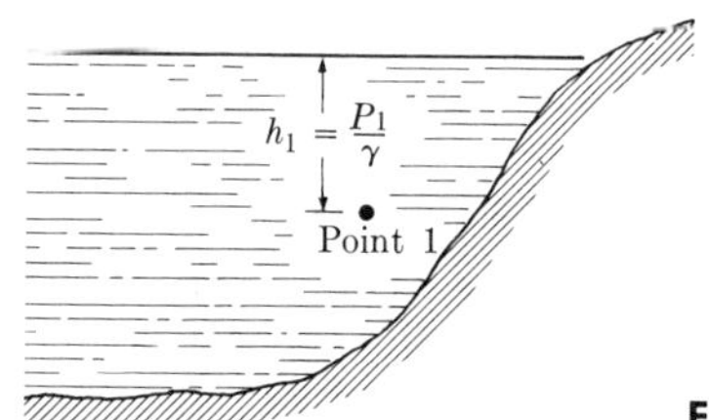

Fig. 3–4

sustain a shear stress, the pressure must act perpendicular to the wall of the pipe. This can be extended to the following generalization:

Pressure always acts perpendicular to the surface restraining the fluid.

In our example, the pressure can be considered to be constant throughout the pipe so long as the pipe diameter is small compared with the head h. If the pipe diameter were large relative to the head, there would be a variation in pressure from top to bottom of the pipe which could not be neglected.

Let us suppose that we insert a flat plate into the pipe (Fig. 3–3). Since the pressure in the pipe is constant and acts perpendicular to the surface, the pressure reaction would be as shown by the arrows in the illustration.

Let us now move from a pipe and column of fluid to an open reservoir, or lake, as shown in Fig. 3–4. Pick any reference point, say Point 1, at some depth, h, in the lake. The pressure at this point would be $P_1 = \gamma h$, which is identical to that at the base of a column of the same liquid of height h. Thus we find that the *head* is independent of the size and shape of the liquid column. Pressure is dependent only upon the head h or the height (depth) of liquid above the point of observation, and upon the density of the fluid.

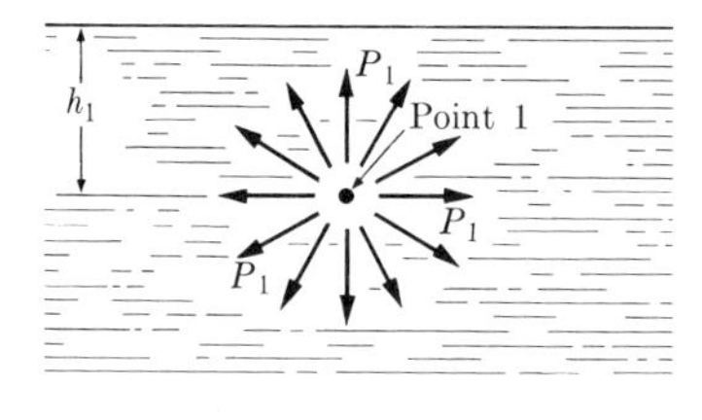

Fig. 3–5

In Fig. 3–5, we take a close look at the reference point, Point 1. The pressure is constant in all directions about the point; that is, the pressure is transmitted undiminished in all directions. All points at the same depth in a static liquid have the same head and are therefore at the same pressure.

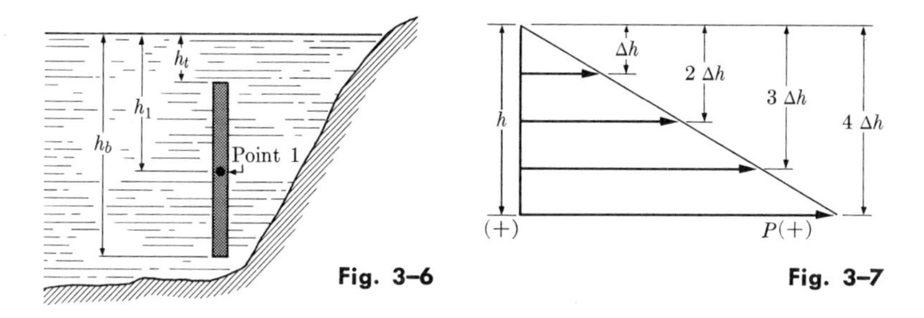

Fig. 3–6

Fig. 3–7

Pressure Distribution with Changing Head

A flat plate is inserted into the lake, as shown in Fig. 3–6. If we allow our reference point to coincide with the center of gravity of the plate, the depth remains h_1. The distance to the top of the plate is h_t, while that to the bottom is h_b. We now have a different problem from that of the point reference; we have a surface of significant area. In determining the pressure reaction on the plate we cannot neglect the variation in head from top to bottom of the plate. The nature of the resulting pressure distribution is shown in Fig. 3–7. We reference head from the surface, considering the downward direction as positive. We can now take increments of depth Δh. Then the pressure at a depth of Δh is $P = \gamma\ \Delta h$; that at $2\ \Delta h$ is $P = 2\gamma\ \Delta h$; etc. We see that pressure variation with depth for an incompressible fluid is linear. This gives us what is called a triangular pressure distribution with depth. Figure 3–8 illustrates how this pressure distribution applies to the submerged plate of our example. If the top edge of the plate were at the surface, $h_t = 0$ and $P_t = 0$ psi. When $h \neq 0$, as in Fig. 3–8, $P_t = \gamma h_t$; obviously, $P_b = \gamma h_b$.

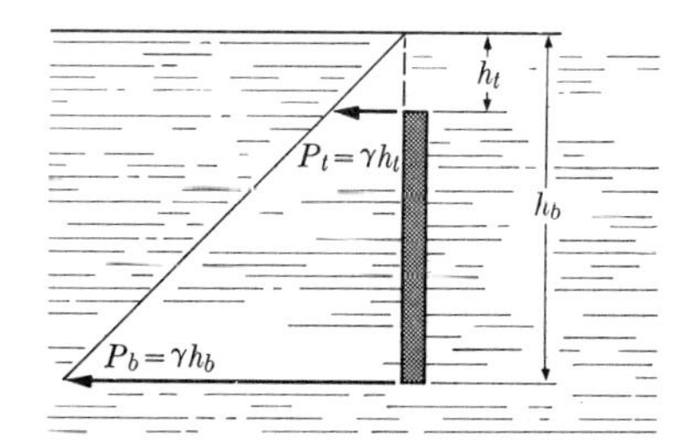

Fig. 3–8

Forces on Submerged Surfaces

What will be the total pressure force on a submerged plate? It will be the sum of all the point pressures acting on all the points of the area of the plate. Mathematically this can be expressed as $F = \int P\, dA$. Solution of this expression* yields the following result: $F = \overline{P} \times A$, where

$F =$ total pressure force,

$\overline{P} =$ pressure at the centroid of the surface,

$A =$ total area of the surface.

Suppose the surface is not perpendicular but makes an angle, θ, with the horizontal (Fig. 3–9). Since the pressure acts perpendicular to the plate, $F = \int \gamma \sin \theta y \, da$. But we can see that $\sin \theta \bar{y} = \bar{h}$, and $\gamma \bar{h} = \bar{p}$, so that the result is $F = \bar{p} \times A$, as before. Thus the resultant total force on a submerged plate, due to pressure, is the *product of the pressure at the centroid of the surface and the total area.*

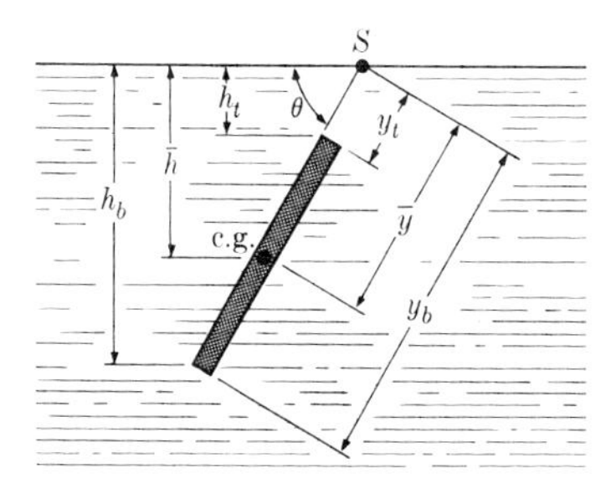

Fig. 3–9

EXAMPLE

Suppose a flat plate 5 ft long and 10 ft wide is submerged in water. The face of the plate makes an angle of 60° with the horizontal. The top edge of the plate is 5 ft below the surface of the water. What is the total pressure force acting on the plate?

Solution.

Step 1. Determine the head at the centroid:

$$\bar{h} = h_t + \tfrac{5}{2} \sin 60° = 5 \text{ ft} + 2.5 \times 0.866 = 7.16 \text{ ft}.$$

Step 2. Calculate the pressure at the centroid:

$$\bar{p} = \gamma \bar{h} = 62.4 \text{ lb/ft}^3 \times 7.16 \text{ ft} = 447 \text{ psf} = 3.1 \text{ psi},$$

or

$$\bar{p} = 0.433 \times S_g \times \bar{h} = 0.433 \times 7.16 \text{ ft} = 3.1 \text{ psi}.$$

* The more advanced student is referred to the complete solution in Appendix A.

Step 3. Calculate the area of the plate:

$$A = l \times w = 5 \text{ ft} \times 10 \text{ ft} = 50 \text{ ft}^2 = 7200 \text{ in}^2.$$

Step 4. Calculate total pressure force:

$$F = \bar{p} \times A = 447 \text{ psf} \times 5 \text{ ft}^2 = 2235 \text{ lb},$$

or

$$F = \bar{p} \times A = 3.1 \times 7200 = 2235 \text{ lb}$$

While the total force due to pressure on a flat surface is equal to the product of the pressure at the centroid and the area, this total force does *not* act at the centroid. It acts at a point called the center of pressure, which lies below the centroid. The distance from the surface of the liquid to the center of pressure is

$$y_p = I/(A \times \bar{y}),^*$$

where

I = the moment of inertia of the area about an axis through point S in Fig. 3–9,
A = area of the surface,
$\bar{y}$ = distance from S to the centroid.

A simplified form of this relationship is:

$$y_p - \bar{y} = k^2/\bar{y},$$

where k = radius of gyration of the plane surface. The difference, $y_p - \bar{y}$, is the distance from the centroid of the plane surface to the center of pressure.

EXAMPLE

Use the previous example for total force and find the location of the center of pressure.

Step 1. Calculate $\bar{y}$:

$$\bar{y} = \frac{5 \text{ ft}}{2} + \frac{5 \text{ ft}}{\sin 60°} = 2.5 \text{ ft} + 5.77 \text{ ft} = 8.27 \text{ ft}.$$

Step 2. For a rectangle of depth d, $k = 0.289d$. Thus

$$k = 0.289 \times 5 = 1.442, \text{ and } k^2 = 2.074.$$

---

* Refer to Appendix A.

Step 3. $y_p - \bar{y} = k^2/8.27 = 2.074/8.27 = 0.251$ ft.

Step 4. $y_p = 8.27$ ft $+ 0.251$ ft $= 8.521$ ft.

Step 5. $h_p = y_p \sin\theta = 8.521 \times 0.866 = 7.4$ ft. Note how this compares to $\bar{h}$ previously calculated ($\bar{h} = 7.16$ ft).

Buoyancy

Another phenomenon related to static pressure is *buoyancy*. By definition, *buoyant force* is the resultant static pressure force exerted on a body wholly or partially submerged in a fluid. About 250 B.C., Archimedes discovered the principle which bears his name; that is,

The buoyant force is equal to the weight of the fluid displaced.

In Fig. 3–10, we observe an object submerged in a fluid. We have already established that the pressure at any point in the fluid is a function of the depth h and the density of the fluid; that is, $p = \gamma \cdot h$. The arrows in Fig. 3–10 represent pressure vectors acting against the surface of the submerged body. It can be demonstrated that $\sum F_x = 0$ and that $\sum F_y \neq 0$. The net force in the y-direction is the buoyant force. It is apparent that the sum of forces acting in the x-direction must be zero; otherwise, the body would move sideways.

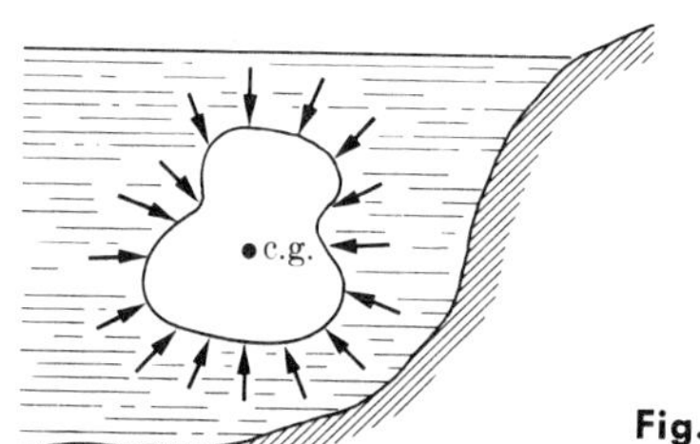

Fig. 3–10

Since pressure is a function of the depth and weight density (specific weight) of the fluid, the actual pressure acting at any point on the body in Fig. 3–10 does not depend upon the body itself. Thus the pressure condition at any point would not change if a body made of a different material were substituted for the original submerged body. Suppose now that such a substitute body were made of the same fluid as the one in which we are immersing it. This substitute body would not move, because its weight is exactly counterbalanced by another force—the weight of the fluid it would displace. If we now switch back to the original body, we would find that it

Fig. 3–11

The ship is a practical application of the buoyancy principles discussed in this chapter. Vessels utilize fluid power to drive and control machinery such as the crane winches, anchor winches, hatch doors, etc. These are applications of Pascal's law and other principles to be covered in later chapters.

would be subject to the same buoyant force, that is, equal to the weight of the fluid displaced. Note that the depth of the submerged object below the surface of the fluid is of no importance here.

When the weight of the fluid displaced equals the weight of the object, the object will float in the fluid. A brief analysis will show that this can only occur when the specific gravity of the immersed object is equal to or less than that of the fluid. If the body's specific gravity is greater than that of the fluid, the weight of the fluid displaced can never equal or exceed the weight of the body. One of the most meaningful applications of the principles of buoyancy is shown in Fig. 3–11.

EXAMPLE

A cast-iron cylinder 12 in. in diameter and 12 in. long is immersed in sea water. What is the buoyant force the sea water exerts on the cylinder?

Solution.

We proceed step by step as follows.

Step 1. Calculate the volume of the cylinder:

$$V = \pi r^2 \times h = \pi \times 6^2 \times 12 = 432\pi \text{ in}^3 = 1357.62 \text{ in}^3.$$

Step 2. Determine the weight of an equal volume of sea water:

$$W_s = \gamma_s V = 1.03 \times 62.4 \text{ lb/ft}^3 \times \frac{1357.62 \text{ in}^3}{1728 \text{ in}^3/\text{ft}^3} = 50.53 \text{ lb.}$$

(Note the dimensional check.)

Step 3. The buoyant force will be equal to this weight of sea water, or $F_b = 50.53$ lb.

If the body had been made of wood, with a specific gravity of 0.75 and with $\gamma_W = 0.027 \text{ lb/in}^3$, what would be the buoyant force in this case?

Step 1. If the entire body were submerged, it would displace 1357.62 in^3 of sea water weighing 50.53 lb.

Step 2. But the piece of wood only weighs

$$W_W = 1357.62 \text{ in}^3 \times 0.027 \text{ lb/in}^3 = 36.66 \text{ lb.}$$

Step 3. Thus the greatest amount of sea water it can displace is 36.66 lb. Obviously, the wooden body must float and be only partially submerged. Hence we have to add the following two steps.

a) Calculate the volume of sea water equal to 36.66 lb:

$$V = \frac{36.66 \text{ lb}}{0.0372 \text{ lb/in}^3} = 985.48 \text{ in}^3.$$

b) Calculate the depth to which the wooden cylinder must sink to displace this much water (assume that the cylinder remains vertical):

$$d = \frac{985.48 \text{ in}^3}{113.10 \text{ in}^2} = 8.45 \text{ in.}$$

IMPORTANT TERMS

Static pressure is the distributed reaction of a mass of fluid on a surface or the force per unit area transmitted through a fluid.

Pressure action is always perpendicular to the surface restraining the fluid.

Head is independent of the shape of the fluid column (container).

Pressure distribution of a fluid of constant density on a submerged surface is linear with depth.

Total pressure force on a submerged surface is equal to the product of the pressure at the centroid and the area of the surface; that is, $F_t = \overline{p} \times A_s$.

Center of pressure is the point at which the total pressure force appears to act; that is, it is the point of application of the resultant pressure force.

Buoyant force is the resultant static pressure force exerted on a body wholly or partially submerged in a fluid. It is equal in magnitude to the weight of the fluid displaced by the body.

PROBLEMS

3–1 What is the pressure at a depth of 100 ft in Lake Michigan?

3–2 What is the pressure at a depth of 150 ft in Long Island Sound?

3–3 In Fig. 3–12, a sluice gate 15 ft long and 10 ft high in a dam is holding back waters on an inland river. Calculate the total force on the gate.

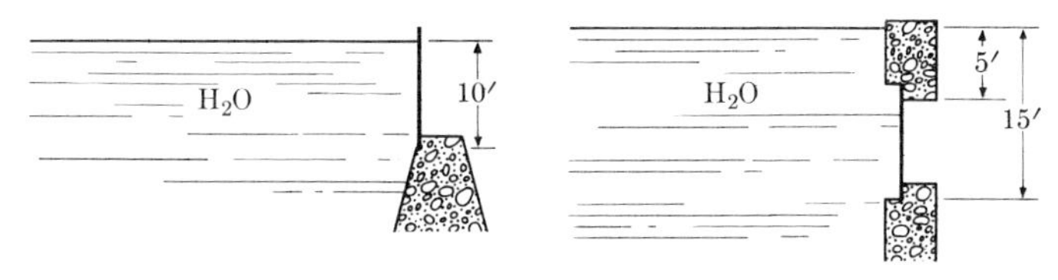

Fig. 3–12 **Fig. 3–13**

3–4 Given the gate of Problem 3–3, what would be the total force if it were 15 ft high and 20 ft long? What is the maximum pressure on the gate?

3–5 A plate covers a clean-out hole in the side of a reservoir, as shown in Fig. 3–13. What is the force on the plate if it is 10 ft wide?

3–6 A rectangular tank filled with hydraulic oil, is 5 ft wide, 3 ft deep, and 10 ft long. Calculate the total force on the sides, ends, and bottom.

3–7 A triangular tank is 15 ft long. The width at the top is 4 ft and the depth is 3 ft. The tank is filled with water. Calculate the forces on the sides and ends.

3–8 A tank 20 ft long has a semicircular vertical cross section of 9.817 ft^2. Calculate the hydraulic forces when the tank is filled with water.

3–9 A closed rectangular tank is 15 ft long, 4 ft wide, and 5 ft deep. It contains water to a depth of 2 ft and oil to a depth of 2 ft. Calculate the hydraulic forces on the ends, sides, and bottom.

3–10 If the tank in Problem 3–9 is pressurized to 2 psi by introducing compressed air into the space above the oil, what will be the forces on the ends, sides, top, and bottom?

3–11 A closed rectangular tank is 5 ft^2 and 3 ft deep. It has a pipe 2 in. in diameter projecting from the top to a height of 15 ft. The tank is filled with water. Hydraulic oil is poured into the pipe to a height of 12 ft. Calculate the pressure forces on the sides, top, and bottom of the tank.

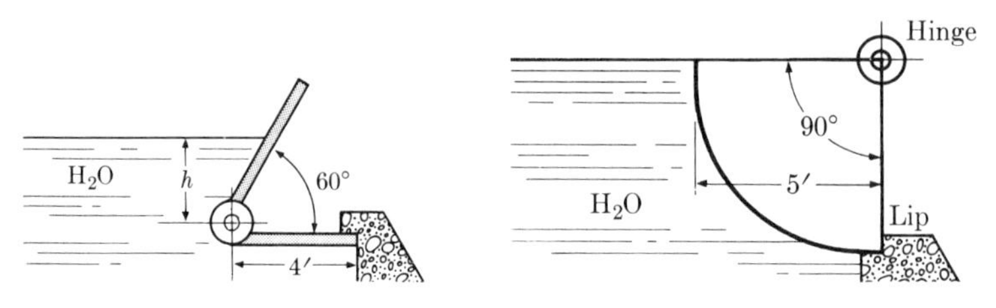

Fig. 3–14 **Fig. 3–15**

3–12 At what depth of water, h, will the hinged gate in Fig. 3–14 open?

3–13 The dam in Fig. 3–15 is 10 ft wide. (a) Is there a depth of water at which the dam gate will open due to pressure forces? (b) What is the force on the lip of the dam when the water level is at the top of the gate?

3–14 An iron casting weighs 150 lb in air and 85 lb in water. What is the volume of the casting?

3–15 A slab of building stone weighs 350 lb in air and 215 lb in water. What is the volume of the slab and the specific gravity of the stone?

3–16 A hollow cylinder made of steel plate is 20 in. in diameter and 10 ft long. If it is made from $\frac{1}{4}$-in. steel (weight = 10.2 lb/ft^2), will it float in water? If so, to what depth will it sink?

3–17 A barge is in essence a rectangular box made of $\frac{1}{4}$-in. steel plate. It is 20 ft wide, 10 ft deep, and 100 ft long. Assume that the stiffening structure adds 10% to the weight of the steel "skin." If minimum free board (projection of the sides out of the water) is 2 ft, how many tons of coal ($S_g = 1.5$) can the barge carry? If the load is changed to wet sand ($S_g = 2.0$), what would be the load in the barge?

3–18 A submarine is a steel tank 25 ft in diameter and 300 ft long. Assume that the outer hull is made of $\frac{3}{8}$-in. steel plate and that the weight of the inner hull and machinery is 250% of the weight of the outer hull. What would the capacity of the ballast tanks have to be in order for the submarine to submerge in the ocean? What, if any, additional capacity would have to be provided to ensure submersion in Lake Michigan?

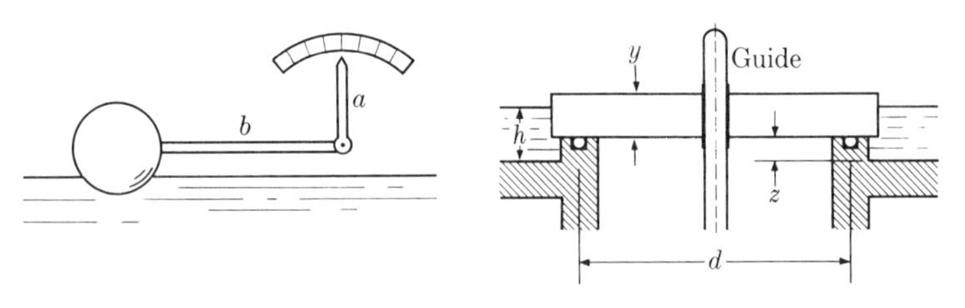

Fig. 3–16 **Fig. 3–17**

3–19 Figure 3–16 shows a spherical weight 10 in. in diameter which is used to sense liquid level in a tank. The sphere is made of 24-gage stainless steel (weight = 1.05 lb/ft^2). If the scale is calibrated for water level in the tank, what error would be introduced if the device were used with hydraulic oil?

3–20 As shown in Fig. 3–17, a cylindrical "can" acts as a valve to close off a drain line. It is made of 20-gage steel sheet (weight = 1.656 lb/ft^2). The cylinder is 12 in. in diameter. What is the relationship between the height y of the cylinder; the depth h of any fluid of specific gravity S_g; z the height of the lip; and ds the diameter of the seal which will cause the valve to open?

CHAPTER 4

Displacement, flow rate, continuity of flow, flow velocity, horsepower

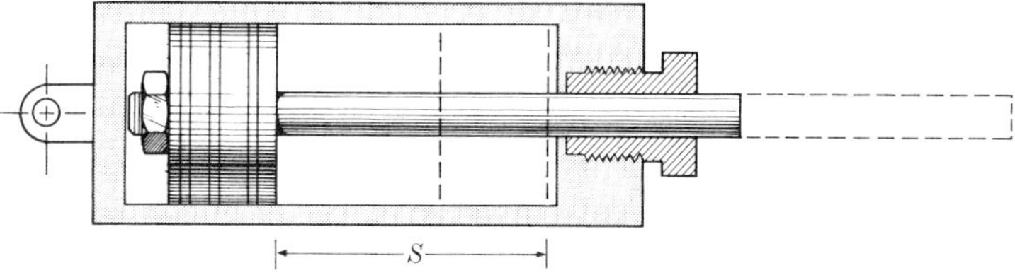

Fig. 4–1

Cylinder Characteristics

Having developed the concept of pressure as the distributed reaction of the load resistance across the area of the piston in the cylinder, it should have become obvious that if this piston is to move down the length of the cylinder, fluid must be injected into the cylinder. In fluid power systems, the device which injects the fluid into the cylinder is called a pump; such is the case in hydraulic systems. In pneumatic systems the device is a compressor.

This distance which the piston moves within the cylinder is the stroke, which is labeled S in Fig. 4–1. When a piston of area A_p moves through a stroke S, the volume swept out in the cylinder is

$$V_d = A_p \times S,$$

where V_d is called the *displacement volume* of the cylinder. It has been general practice in industry to express displacement volumes in units of cubic inches. Thus the area must be in square inches and the stroke in inches.

If it takes a time t for the piston to move through its stroke, the rate of displacement will be $V_d/t = A_p \times S/t$. Since $S/t =$ velocity v, then $V_d/t = A_p \times v$. We can also think of this as the rate at which fluid must flow into the cylinder. Flow rate is designated as Q:

$$Q = \frac{V_d}{t} = A_p \times v.$$

Note that displacement is a finite volume, or quantity, of fluid, which is expressed in units of cubic feet or cubic inches in the English system of measurement; cubic centimeters are generally used in the metric system.

On the other hand, flow rate is a time-based quantity—a rate. Its usual units are cubic feet per second (cfs) or cubic inches per second (cis) in the English system, and cubic centimeters per second (ccs) or liters per second in the metric system.

Because of the confusion of units in the English system, it has become general usage in industry to state *displacement* in cubic inches or cubic feet and *flow rate* in gallons per minute (gpm). The following conversions will be of value in future problem solving:

$$1 \text{ gal} = 231 \text{ in}^3 = 8.345 \text{ lb (water)},$$
$$1 \text{ ft}^3 = 7.48 \text{ gal} = 62.4 \text{ lb (water)}.$$

Flow Velocity

If we look again at the piston in the cylinder of Fig. 4–1, we can consider it as a surface of area A_p moving along the cylinder bore with a velocity $v = S/t$. In order for the incoming fluid to push the piston along at this velocity, the fluid itself must have an average velocity equal to v. The flow rate equation, $Q = A_p \times v$, expresses the relationship between flow rate, the area through which the flow is occurring, and the average velocity. When we solve this equation for flow velocity, we get the expression $v = Q/A_p$.

Let us consider the pipe through which the fluid must flow in order to get to the cylinder. Suppose the pipe has a cross-sectional area of A_0, square units (inches or feet). An important point here is that the flow rate in the pipe must be the same as the flow rate in the cylinder itself. This follows from the acknowledgement of the fact that unless fluid is taken away or added to the flowing system, the rate of flow for incompressible

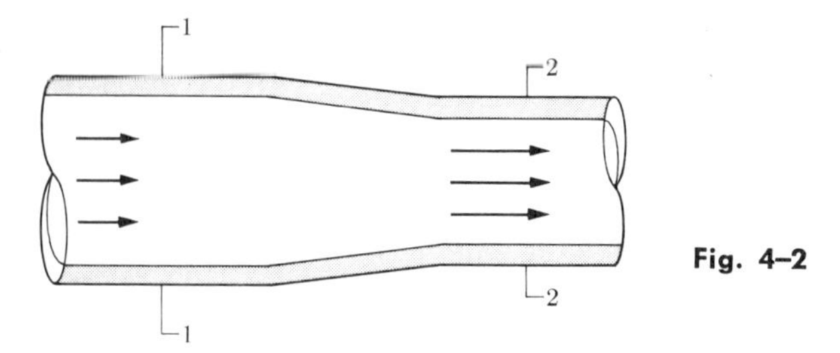

Fig. 4–2

flow conditions is constant throughout the system. Thus the flow rate in the connecting pipe would be $Q = A_0 \times v_0$, where v_0 is the average velocity in the pipe.

Since Q is constant we can equate the two expressions:

$$Q = A_p \times v = A_0 \times v_0. \tag{4–1}$$

We see that the velocities are not equal but are inversely proportional to the cross-sectional area of the fluid conductor. This expression is a simple statement of the principle of *continuity of flow*. The illustration of Fig. 4–2 may make it easier to visualize continuity of flow. The area at section 1 is larger than that at section 2. The continuity equation $Q = A_1 v_1 = A_2 v_2$ tells us that $v_1 < v_2$, because $A_1 > A_2$.

EXAMPLE

In a pipe with an area of 4 in^2, water is flowing at a rate of 20 gpm. The pipe makes a transition to one of cross-sectional area equal to 1.5 in^2. What is the flow velocity in each of the sections?

Solution.

Step 1. Convert flow rate to units consistent with the areas:

$$20 \text{ gpm} \times 231 \text{ in}^3/\text{gal} = 4620 \text{ in}^3/\text{min}.$$

Step 2. From the continuity equation $Q = A_1 v_1 = A_2 v_2$, solve for the velocities:

$$v_1 = \frac{Q}{A_1} = \frac{4620 \text{ in}^3/\text{min}}{4 \text{ in}^2} = 1155 \text{ in./min},$$

$$v_2 = \frac{Q}{A_2} = \frac{4620 \text{ in}^3/\text{min}}{1.5 \text{ in}^2} = 3080 \text{ in./min}.$$

Let us not lose sight of the fact that we have been developing these ideas on the basis of our "ideal liquid," as defined in Chapter 1.

Mass Flow Rate

So far we have dealt with flow rate in terms of volume per unit time. It is also possible to consider it from the viewpoint of weight per unit time or mass per unit time. In this latter light we speak of the *mass flow rate.* In normal usage, when we speak of *flow rate* we mean volume flow rate; we use the term *mass flow rate* to denote the rate of mass transfer through the conductor.

Thus if Q is the symbol for the volume flow rate, then γQ is the weight flow rate, where γ is the specific weight, and ρQ is the mass flow rate, where ρ is the density.

Units for γQ are pounds per unit time (seconds or minutes); units for ρQ are slugs per unit time, provided Q is expressed in the same basic unit of volume as is γ or ρ.

Work Done in Pumping a Fluid

Now that we have established the concepts of equivalency of flow rate, mass flow rate, and pressure and head, we are in a position to consider the work done in transferring a liquid through a conductor under a change in pressure. We all know intuitively or by observation that it takes power to pump a liquid. An electric motor or internal combustion engine drives a pump. Liquid enters the pump on the inlet side at a pressure approximating atmospheric pressure (0 psig) and exits from the pump on the outlet side at some elevated discharge pressure P_0. Something occurs in the pump which results in work being done on the fluid. The occurrence depends on the equivalency of the mass flow rate and of pressure and head. See Fig. 3–2. Pumping a liquid through a pump at a rate of γQ lb/min across a pressure differential of $P_0 - 0$ psig requires the same amount of *work* done on the liquid as raising a weight of liquid equal to γQ lb to a height of

$$h = P_0/\gamma \text{ ft (or in.)}.$$

The result in either case is ft-lb of work being done.

EXAMPLE

Assume a pump is supplying water to a cylinder at a rate of 1 gpm. The discharge pressure is 1000 psi. How much work is done in one minute?

Solution.

Step 1. $Q = 1 \text{ gpm} = \dfrac{231 \text{ in}^3\text{/gal}}{1728 \text{ in}^3\text{/ft}^3} \times 62.4 \text{ lb/ft}^3 = 8.35 \text{ lb/min.}$

Step 2. $h = \dfrac{P_0}{\gamma} = \dfrac{1000 \text{ psi} \times 144 \text{ in}^2/\text{ft}^2}{62.4 \text{ lb/ft}^3} = 2320$ ft of head.

Step 3. Work $= G = 8.351$ lb $\times$ 2320 ft $=$ 19,300 ft-lb.

To ensure consistency of units in these calculations, observe the responsibility placed on the student by the English system of measures. Only a dimensional check will assure the proper answer.

Horsepower Required to Pump Fluids

The development of the *hydraulic horsepower formula* is an extension of the ideas discussed above.

Horsepower is the rate of doing work or the rate of energy transfer.

$$\begin{aligned} 1 \text{ hp} &= 550 \text{ ft-lb/sec} = 33{,}000 \text{ ft-lb/min} \\ &= 42.44 \text{ Btu/min} = 745.7 \text{ watts } (=1.014 \text{ metric hp}). \end{aligned}$$

In the example above, the flow rate Q was given in gpm and the pressure in psi. In order to develop a hydraulic horsepower formula, it is necessary to reduce these quantities to units which when multiplied will yield ft-lb/time. Thus

$$Q_{\text{gpm}} \times \frac{231 \text{ in}^3/\text{gal}}{1728 \text{ in}^3/\text{ft}^3} \times 62.4 \text{ lb/ft}^3$$

yields lb/min, and

$$P_{\text{psi}} \times \frac{144 \text{ in}^2/\text{ft}^2}{62.4 \text{ lb/ft}^3}$$

yields ft of head; then

$$P \times Q = \frac{Q \times 231 \times 62.4 \times p \times 144}{1728 \times 62.4} \text{ ft-lb/min},$$

$$\text{hp} = \frac{Q \times 231 \times 62.4 \times P \times 144}{1728 \times 62.4} \times \frac{1}{33{,}000},$$

$$\text{hp} = \frac{P \times Q}{1714}, \tag{4–2}$$

where P is in psi and Q is in gpm. The horsepower calculated by the above

Fig. 4–3

The principles of displacement, flow rate, force, velocity, and hydraulic horsepower, covered in this and previous chapters, are all part of the problems which face the engineer when he attempts the design of a modern, complex construction machine, such as this single-pass paving machine. It has been stated that the construction equipment industry, as we know it today, owes its existence to fluid power. In no other way would it be practical to transmit the high power required by these machines to the remote parts of the vehicle. Control is also greatly simplified.

formula would represent the rate of energy transfer corresponding to a rate of fluid flow, Q, through a pressure difference P.

If we consider a cylinder which has a piston area A_p in^2 and a pressure P psi acting on the piston, the force pushing on the piston rod will be $F_r = P \times A_p$ lb. See Fig. 4–1. The rod will move through a stroke S in. (or ft). Thus the work done by the cylinder is $G = F_r \times S$ in.-lb (or ft-lb).

If the cylinder moves through the stroke in time t, the rate of doing work is

$$F_r \times (S/t) \text{ in-lb/sec or min (or ft-lb/sec or min).}$$

If we divide the above expression by 550 ft-lb/sec/hp when t is in seconds, or by 33,000 ft-lb/min/hp when t is in minutes, we will determine the cylinder's horsepower. Assuming, for the moment, 100% efficiency, this cylinder horsepower will equal the hydraulic horsepower calculated by Eq. (4–2). Figure 4–3 illustrates a piece of modern construction equipment which makes use of the principles outlined in this chapter.

IMPORTANT TERMS

Stroke is the distance through which the piston-rod assembly of a cylinder moves.

Displacement volume for a cylinder is the product of the area of the piston and the stroke: $V_d = A_p \times S$.

Flow rate (volume flow rate) is the quantity (volume measure) of fluid which passes a reference point per unit time. *Mass flow rate* is the quantity (mass) of fluid which passes a point per unit time; the units are expressed in slugs/sec or min. *Weight flow rate* is the quantity (weight) of fluid passing a point in unit time; units are expressed in lb/sec or min.

Flow velocity is the velocity of the fluid stream passing a given point. Average velocity is equal to the flow rate Q divided by the cross-sectional area of the flowing stream A, namely, $\bar{v} = Q/A$ ips or fps.

Continuity of flow is the principle which states that the flow rate in a continuous stream is constant: $Q = A_1v_1 = A_2v_2 = K$.

Work is the product of a force and the distance through which it moves. It represents a change in energy condition of a system.

Hydraulic horsepower is the horsepower required to move a fluid through a pressure difference P, at a flow rate Q: hp $= P \times Q/1714$ where $P =$ psi, $Q =$ gpm.

PROBLEMS

4–1 In the cylinder in Fig. 4–4, $A_1 = 15$ in^2 and the stroke $S = 30$ in. What is the displacement on the blank end?

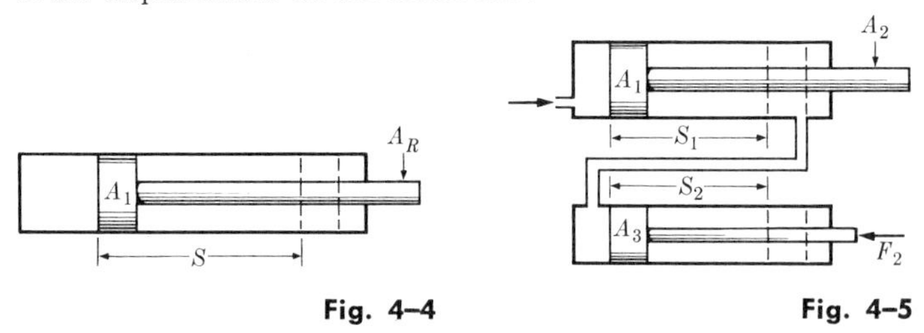

Fig. 4–4

Fig. 4–5

4–2 If $A_1 = 20$ in^2 in Fig. 4–4, what is the displacement (blank end) per inch of stroke? How many gallons would the cylinder hold if $S = 22$ in.?

4–3 If the rod area $A_R = 5$ in^2 in Fig. 4–4, what is the displacement of the rod end of the cylinder?

4–4 How do you account for the difference between the answers to Problems 4–1 and 4–3?

4–5 Figure 4–5 shows two cylinders in series: $A_1 = 20$ in^2, $A_2 = 5$ in^2, $A_3 = 12$ in^2. If $S_1 = 30$ in., what would the maximum distance be for S_2?

4–6 In Fig. 4–5, assume that we push on the piston rod of the second cylinder with a force F_2. If we move the second piston through a distance ($S_2 = -20$ in.), how far (S_1) will the first piston move?

4–7 How might the system of Fig. 4–5 be designed so that *both* pistons moved the same distance when the first piston was extended? Under these conditions, would they both retract the same distance when F_2 pushed the second piston back into its cylinder barrel?

4–8 Under the conditions of Problem 4–1, assume that it takes 5 sec for the piston to move through its stroke. At what *rate* would fluid have to flow into the blank end of the cylinder to keep it filled?

4–9 We know from practical experience that under sufficient pressure to overcome the load on the rod, it is the fluid flowing into the cylinder which causes the piston to move. Using the data of Problems 4–1 and 4–8, calculate the piston velocity in the cylinder. How would this relate to the fluid velocity in the cylinder?

4–10 Under the conditions of Problem 4–8, assume that the fluid is flowing to the cylinder through a tube 1 in. in diameter. What is the fluid velocity in the tube?

4–11 Why is the velocity calculated in Problem 4–9 different from that calculated in Problem 4–10? What principle describes this phenomenon? Write the mathematical expression for this principle.

4–12 Applying the principle described in the answer to Problem 4–11, use the series cylinder system of Problem 4–5. Assume that piston 1 moves through stroke S_1 in 3 sec. Fluid enters cylinder 1 through a nominal commercial pipe $\frac{3}{4}$ in. in diameter. Cylinders 1 and 2 are connected by tubing having a 1-in. internal diameter. Calculate the flow rates and flow velocities in all parts of the system.

4–13 In Problem 4–12, suppose that the cylinders are connected by a $1\frac{1}{2}$-in. commercial pipe. What are the flow rate and flow velocity in the $1\frac{1}{2}$-in. pipe?

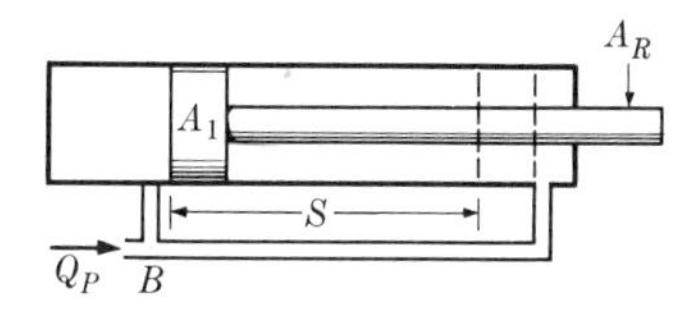

Fig. 4–6

4–14 Figure 4–6 shows a regenerative cylinder, as described in Chapter 2. Assume that $A_1 = 25$ in^2, $A_R = 7$ in^2, $S_1 = 36$ in. If it takes 3 sec to move the piston through the full stroke, what must the input flow rate Q_p be? [*Hint:* Take the sum of the flows into the junction at B and equate them to the sum of the flows leaving the junction.]

4–15 Water is flowing through a pipe 6 in. in diameter at a flow velocity of 20 ft/sec. The pipe diameter reduces to 3 in. What are the flow rate and the flow velocity in the reduced section?

4–16 Oil flows through a pipe 1 in. in diameter at a rate of 20 gpm. A valve placed in the pipe has a passage in it of $\frac{3}{4}$ in. in diameter. What are the flow velocities in the pipe and through the valve?

4–17 The cylinder in Fig. 4–7 is called "double acting"; that is, fluid can be introduced at *either* end to move the piston rod in either direction. Use the flow rate calculated in Problem 4–8 (let $A_R = 7$ in^2), but introduce the fluid into the rod end instead of the blank end. What is the flow rate Q_{out} of fluid out of the port in the blank end?

Fig. 4–7 **Fig. 4–8**

4–18 Fluid is introduced into the pump in Fig. 4–8 through a $1\frac{1}{2}$-in. pipe port, and discharged through a 1-in. pipe port. If the discharge flow rate Q_{out} is 40 gpm, what are the flow velocities in the inlet and outlet ports?

4–19 In the cylinder of Problem 4–1, how much work is done by the cylinder (neglect losses) if the pressure in the blank end is 1000 psi?

4–20 In the regenerative cylinder of Problem 4–14, how much work is done if the pressure is 2000 psi? How much horsepower is developed?

4–21 Use the cylinder system in Fig. 4–5. Assume that $F_2 = 6000$ lb. If the pressure in the blank end of cylinder 1 is $P_1 = 1500$ psi, how much work is done by the system when $S_1 = 20$ in.?

4–22 Under the conditions of Problems 4–12 and 4–21, calculate the hydraulic horsepower which would have to be delivered by the pump supplying fluid to the system.

4–23 If the discharge pressure of the pump in Problem 4–18 is 1500 psi, calculate the hydraulic horsepower delivered to the system.

4–24 A standpipe 50 ft high rests on the top of a hill 135 ft above the level of a pumping station. The pump sits in a pit 10 ft below floor level. If it delivers 10,000 gpm to the standpipe, what is the efficiency of the pump if it must be driven by an electric motor of 600 hp?

4–25 The ballast pumps in a submarine submerged at a depth of 250 ft discharge 35,000 gpm of sea water. Neglecting losses, determine the horsepower required to drive the pumps.

4–26 The fuel pumps in a ballistic missile deliver 3000 lb/min of liquid fuel ($S_g = 1.47$) to the rocket engine nozzles at a pressure of 250 psi. What horsepower must be delivered to the pumps?

CHAPTER 5

Conservation of energy, Bernoulli's equation

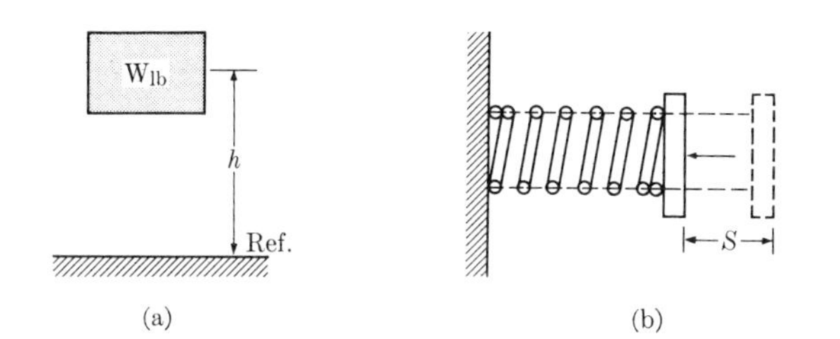

Fig. 5–1

Energy Relationships in Fluid Systems

You are undoubtedly familiar with the fact that there are two basic energy states for Newtonian systems, that is, systems which are massive enough to obey Newton's laws of motion.

The first of these states is *potential energy*, which is regarded as stored energy or, in a system, as the ability to do work. Simple examples are given in Fig. 5–1. Figure 5–1(a) shows a weight W at an elevation h relative to a reference. The weight has potential energy $W \cdot h$ in relation to the reference plane. Figure 5–1(b) shows a spring with a spring constant, k (lb/in). When the spring is compressed a distance s in., the potential energy stored in the spring is $k \times s/2$.

The second energy state is *kinetic energy*, which is determined by the motion or velocity of a body:

$$\text{KE} = \tfrac{1}{2}Mv^2$$

where M is the mass of the body and v is its velocity in feet per second.

In a Newtonian system, the principle of conservation of energy states that the total energy in the system remains constant. Thus a change in potential energy level presumes a corresponding change in kinetic energy: $\Delta PE = \Delta KE$. Or, as potential (stored) energy passes from the stored state to a condition of doing work, it is converted to kinetic energy. Thus, if the weight shown in Fig. 5–1(a) were to fall from its elevation h, all of its potential energy, $W \cdot h$, would be converted to kinetic energy, $\frac{1}{2}Mv^2$, by the time the weight reached the reference plane.

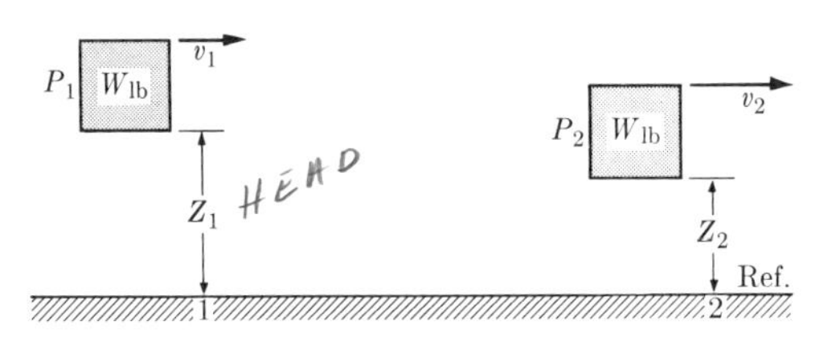

Fig. 5–2

Consider now Fig. 5–2. At point 1 we have a quantity of a fluid, W lb, which is moving at an average velocity v_1, has been raised to a pressure P_1 psi, and is at an elevation Z_1 relative to the reference plane. The total energy content of the fluid quantity at point 1 is

$$E_1 = \text{PE(elevation)} + \text{PE(pressure)} + \text{KE}$$
$$= WZ_1 + W\frac{P_1}{\gamma} + \frac{1}{2}\frac{Wv_1^2}{g}.$$

Note that the units for each term are foot-pounds or inch-pounds, which you will recognize as the usual energy units for the English system of measure.

At point 2 in Fig. 5–2, the quantity of fluid has a different elevation, pressure, and velocity, indicating that it has undergone some changes in condition between point 1 and point 2. The expression for total energy at point 2 becomes

$$E_2 = WZ_2 + W\frac{P_2}{\gamma} + \frac{1}{2}\cdot\frac{W}{g}\cdot v_2^2.$$

Bernoulli's Equation

The principle of energy conservation tells us that for ideal fluids, the total energy in the system is constant. Thus we can equate these two expressions for energy:

$$WZ_1 + WP_1/\gamma + (\tfrac{1}{2})(Wv_1^2/g) = WZ_2 + WP_2/\gamma + (\tfrac{1}{2})(Wv_2^2/g).$$

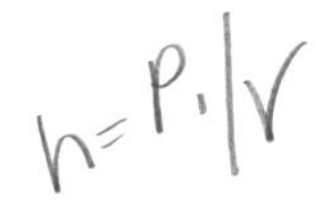

The units for each term are foot-pounds (or inch-pounds). Note that W, the weight of our hypothetical quantity of fluid, is common to each term. We can divide by W (lb), and arrive at the following expression:

$$Z_1 + P_1/\gamma + v_1^2/2g = Z_2 + P_2/\gamma + v_2^2/2g, \tag{5-1}$$

where each term in Eq. (5–1) represents the energy, in foot pounds or inch pounds, *per pound* of *fluid flowing*. Note that the units for each term are now feet or inches.

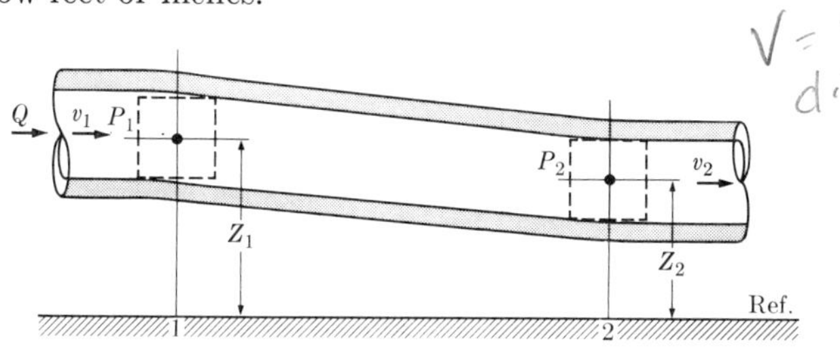

Fig. 5–3

Since all terms in (5–1) are given in units of *head*, feet or inches, they are called, respectively, *elevation head*, *pressure head*, and *velocity head*. Elevation head and pressure head are related to the potential energy of the stream, while velocity head is a function of the kinetic energy of the flowing stream.

Equation (5–1) is known as *Bernoulli's equation* and is one of the concepts fundamental to the fluid mechanics of incompressible fluids.

It is perhaps a little difficult to visualize abstractly a quantity of fluid in space, such as is illustrated in Fig. 5–2. If we superimpose a conductor, or pipe (Fig. 5–3), the picture becomes more realistic. We can now think of the quantity as a unit volume of fluid at point 1. At some time later the same unit has moved along the pipe to point 2. The energy components will be the same as they were previously, and Bernoulli's equation can be derived in a similar manner. This concept is basic to our study of fluid mechanics.

EXAMPLE

Consider a simple flow system (Fig. 5–4) consisting of a conductor, or pipe, of varying cross-sectional area. Assume that the flow rate $Q = 1000$ gpm, $P_1 = 1000$ psi, $A_1 = 2$ in^2, and $A_2 = 1$ in^2. What is P_2?

From the principle of flow continuity, we recognize that Q is constant, that is:

$$Q = A_1 v_1 = A_2 v_2.$$

The next step will be to write Bernoulli's equation and determine the known

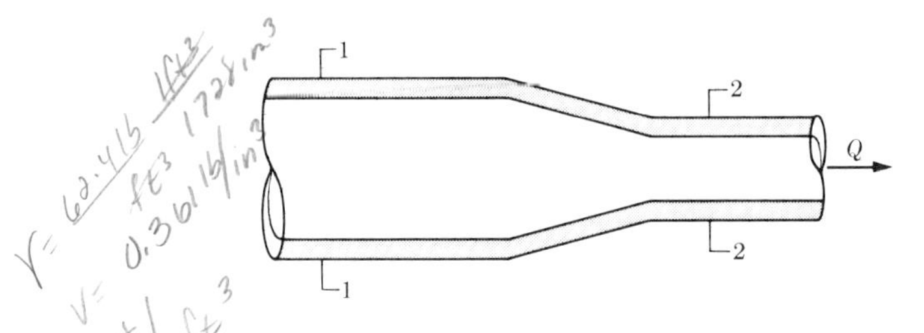

Fig. 5–4

and unknown quantities:

$$P_1/\gamma + v_1^2/2g + Z_1 = P_2/\gamma + v_2^2/2g + Z_2.$$

Substituting, we obtain

$$1000(\text{lb/in}^2)/0.0361(\text{lb/in}^3) + v_1^2/772 = p_2/0.0361(\text{lb/in})^3 + v_2^2/772.$$

Note that $Z_1 = Z_2$ for a horizontal pipe, as shown in Fig. 5–4. We see that we must determine the velocities in order to find the desired unknown, P_2. Thus

$$v_1 = \frac{Q}{A_1} = \frac{1000 \text{ gal/min} \times 231 \text{ in}^3/\text{gal}}{2 \text{ in}^2 \times 60 \text{ sec/min}}$$
$$= 1925 \text{ ips},$$

and

$$v_2 = \frac{Q}{A_2} = \frac{1000 \text{ gal/min} \times 231 \text{ in}^3/\text{gal}}{1 \text{ in}^2 \times 60 \text{ sec/min}}$$
$$= 3850 \text{ ips}.$$

Note that the velocity calculations were made to the time base of seconds since the acceleration due to gravity, g, is based on seconds. Substituting, we find

$$\frac{1000}{0.0361} + \frac{(1925)^2}{772} = \frac{P_2}{0.0361} + \frac{(3850)^2}{772},$$

from which

$$P_2 = 470 \text{ psi}.$$

Ideal Flow Versus Actual Flow Conditions

The above example points out some facts of interest.

1) For a constant flow rate Q, when the cross-sectional area of the conductor decreases from one point to another, the flow velocity increases, and vice versa.

2) The velocity head varies as the square of the flow velocity.

3) When the flow velocity increases, there is a corresponding decrease in pressure; conversely, when there is a decrease in velocity, there is an increase in pressure.

In Chapter 1, we stated that initially we would discuss only ideal fluids, and that the losses which occur in actual (nonideal) fluids would be considered later as "corrections" to the ideal case. It is still premature to discuss the calculation of such losses, but we must recognize their existence now in order to make Bernoulli's equation completely correct.

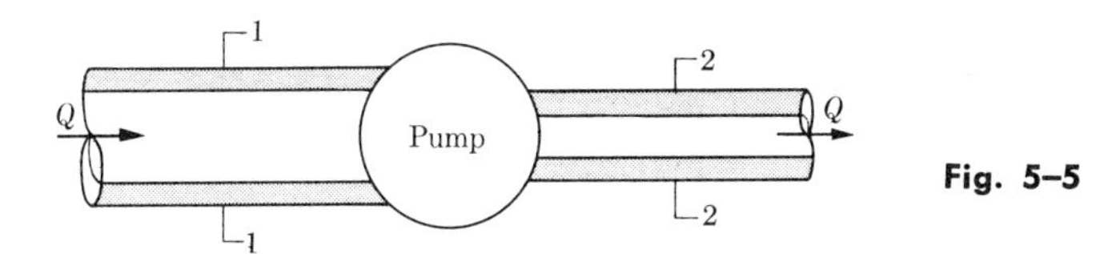

Fig. 5–5

For this purpose, all losses in a flow system will be combined in one *head loss* which we will designate as H_L. Furthermore we must recognize that it is possible to add energy to a system between the two reference points, for example, by means of a pump. (See Fig. 5–5.) The energy so added will be potential energy in the form of increased pressure head. Such head additions will be lumped into one pump head term, H_p. The complete form of Bernoulli's equation will then be

$$\frac{P_1}{\gamma} + \frac{v_1^2}{2g} + Z_1 + H_p = \frac{P_2}{\gamma} + \frac{v_2^2}{2g} + Z_2 + H_L. \tag{5–2}$$

Note that the head increase due to a pump placed between points 1 and 2 is always added to the left-hand side of the equation. The head loss, H_L, is always added to the right-hand side of the equation, as shown in Eq. (5–2).

Energy Diagram

The reason for the procedure described above can be illustrated graphically by means of an *energy diagram* (see Fig. 5–6). The vertical line, labeled 1 on the diagram, represents reference point 1 in the flowing stream. (See Fig. 5–4.) The various heads (pressure, velocity, elevation) representing energy are plotted vertically along this line.

The vertical line to the right side of the diagram, labeled 2, represents reference point 2 in the flowing stream. As was the case at point 1, the heads are plotted along this line.

We have established that there will be some energy losses in this flowing stream. Since the velocities must obey the principle of flow continuity, the loss cannot show up as a loss in velocity head. Since the pipe in our example is horizontal, there is no change in elevation head. Therefore it follows that the loss must be a loss in pressure. Thus the pressure head, P_2/γ, will actually be less than what we would have calculated for the ideal case, using Eq. (5–1). The difference is H_L and is plotted at point 2 on our diagram to bring the total energy content up to the constant energy line. When we write the right-hand side of Bernoulli's equation, corresponding to point 2, we include *all* the terms, including H_L, plotted at point 2; thus $P_2/\gamma + v_2^2/2g + Z_2 + H_L$.

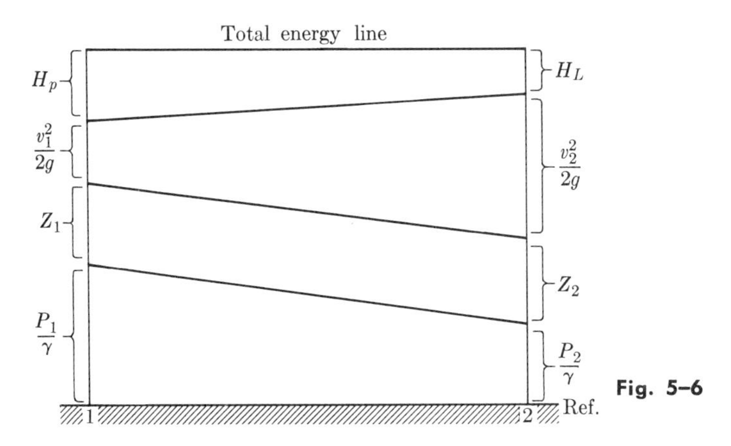

Fig. 5–6

We stated that energy was to be added to the system by means of a pump placed between reference points 1 and 2 (Fig. 5–5). The effect of this step is to raise the level of the constant energy line beyond point 1. The head corresponding to this increase in energy, H_p, must be added to the left-hand side, or at point 1, in order to bring the sum of the energy components up to the new constant line for the diagram. Thus

$$\frac{P_1}{\gamma} + \frac{v_1^2}{2g} + Z_1 + H_p.$$

Note that when we write Bernoulli's equation we consider two reference points only. With the exception of a pump head addition, there are no terms in the equation resulting from intermediate points. Head loss always occurs in the direction of flow, that is, from the upstream point (point 1) to the downstream point (point 2).

Bernoulli's equation is basic to the analysis of many of the fluid systems and devices which we will consider later. Typical of these is the articulated arm shown in Fig. 5–7.

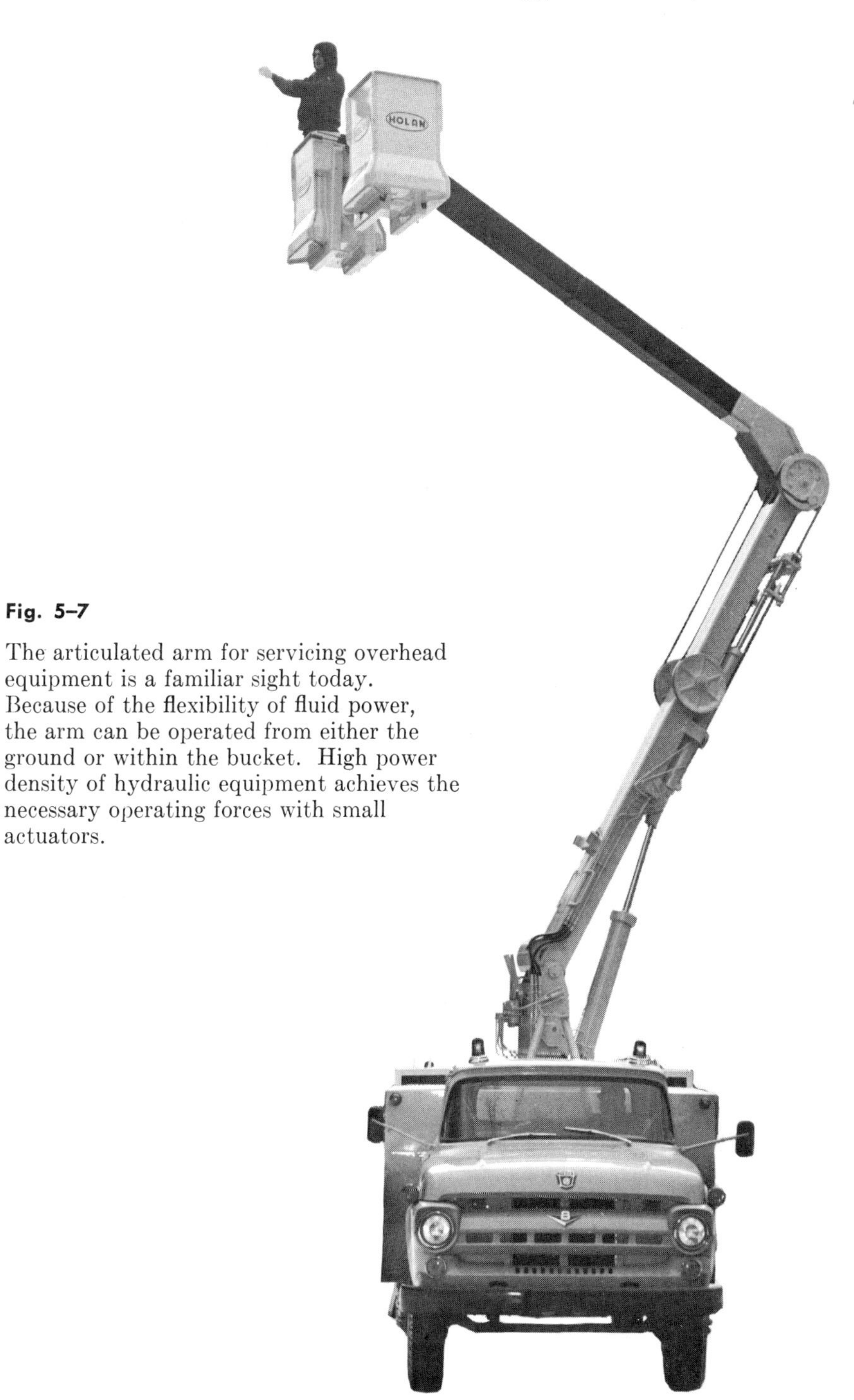

Fig. 5–7

The articulated arm for servicing overhead equipment is a familiar sight today. Because of the flexibility of fluid power, the arm can be operated from either the ground or within the bucket. High power density of hydraulic equipment achieves the necessary operating forces with small actuators.

IMPORTANT TERMS

Potential energy in a fluid system has two components: (1) potential energy due to elevation; (2) potential energy due to pressure.

Kinetic energy in a fluid system is a function of the mass flow rate and the velocity of the flowing stream of fluid.

Bernoulli's equation is the mathematical expression of the principle that the energy content of a flowing fluid system is constant:

$$P_1/\gamma + v_1^2/2g + Z_1 + H_p = P_2/\gamma + v_2^2/2g + Z_2 + H_L.$$

Head loss, H_L, is the reduction in pressure head in a fluid system due to viscous, throttling, and turbulent losses induced by the motion of the fluid.

Head added by a pump, H_p, is the increase in pressure head due to the energy added to the system.

PROBLEMS

5–1 A section of a pipe system is shown in Fig. 5–8, in which $Q = 600$ gpm of water. (a) What is the kinetic energy at sections 1–1 and 2–2? (b) What are the velocity heads at 1–1 and 2–2? (c) What is the pressure differential between sections 1–1 and 2–2?

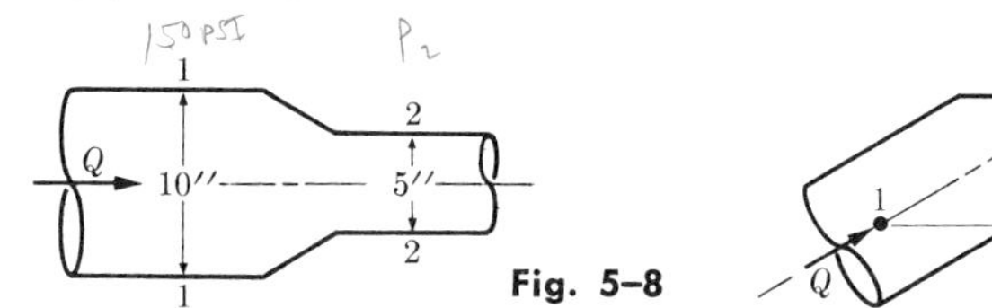

Fig. 5–8

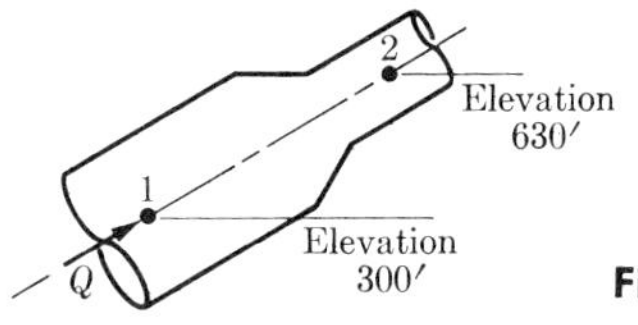

Fig. 5–9

5–2 In a flow conductor similar to that of Problem 5–1, the diameter at section 1–1 = 4 in., and the diameter at 2–2 = 2 in. The pressure at section 1–1 is 150 psi; $Q = 200$ gpm of hydraulic oil.* Calculate the pressure at section 2–2.

5–3 In the conductor in Fig. 5–9 the cross-sectional area at point 1 is 10 in^2, and that at point 2 is 4 in^2. The flow rate $Q = 150$ gpm of sea water. What is the pressure difference between points 1 and 2?

5–4 In Problem 5–3, if the head loss between points 1 and 2 is 30 ft, what is the pressure difference?

5–5 In a system similar to that of Problem 5–4, a pump is placed in the line between points 1 and 2 at an elevation of 500 ft. What horsepower would

---

* Refer to Table B–1, p. 219. Use "mineral oil."

the pump have to deliver for the pressure at point 2 to equal the pressure at point 1?

5–6 What is the significance of each term in the Bernoulli equation?

5–7 In the pump in Fig. 5–10, $Q_{out} = 35$ gpm. What is Q_{in}? What is the pressure difference between points A and B? Assume fluid is water.

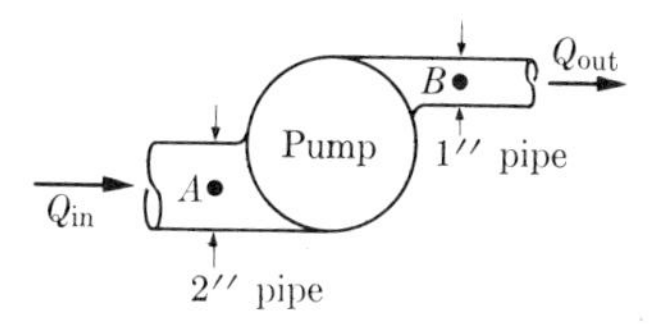

Fig. 5–10

5–8 If the input to the pump in Problem 5–7 is 5 hp, what is the pressure difference between points 1 and 2? What does this pressure difference indicate?

5–9 A pumping system is shown in Fig. 5–11, in which $Z_1 = 100$ ft, $Z_2 = 110$ ft $= Z_3$, and $V_3 = 20$ fps. An electric motor delivers 25 hp to the pump, which is 75% efficient. $H_{L_{1-2}} = 2$ ft; $H_{L_{3-4}} = 10$ ft. To what height, Z_4, could water be pumped? What would the pressure be at point 2?

5–10 In the system of Problem 5–9, assume that $Z_4 - Z_3 = 22.2$ ft, $H_{L_{3-4}} = 1$ ft. With other conditions the same as in 5–9, what would be the maximum theoretical suction lift, $Z_2 - Z_1$?

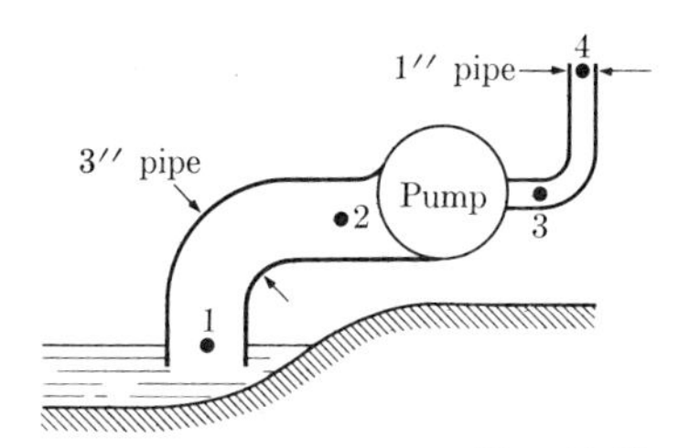

Fig. 5–11

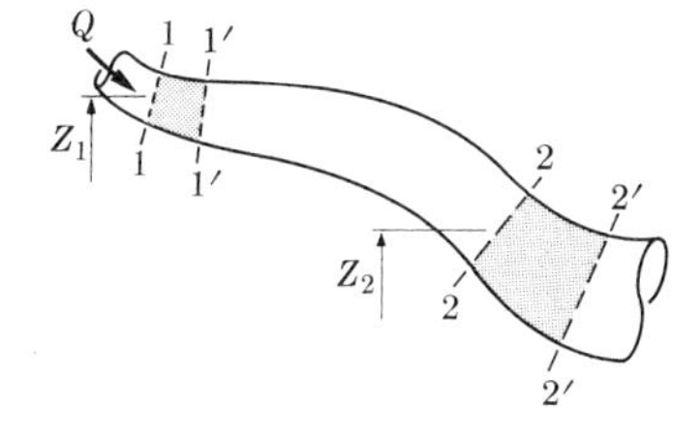

Fig. 5–12

5–11 Figure 5–12 shows a flow conductor of variable cross-sectional area and varying elevation along its length. Using work-energy relationships, derive Bernoulli's equation. [*Hint:* Consider that in the time interval dt, the fluid will have moved from section 1–1 to 1′–1′ with velocity v_1, and similarly from 2–2 to 2′–2′ with velocity v_2. Determine the pressure forces and the distance through which the fluid moved in time dt = work done. Furthermore, consider that, because of flow continuity, the volume of fluid bounded by 1–1 and 1′–1′ equals the volume of fluid bounded by 2–2 and 2′–2′. Thus the work done by gravity on an incremental mass is the product of that mass and the difference in elevations between points 1 and 2. The change in kinetic energy can be determined from the incremental mass and change in velocity between points 1 and 2. Then equate work done to Δ KE.

CHAPTER 6

Applications of Bernoulli's equation

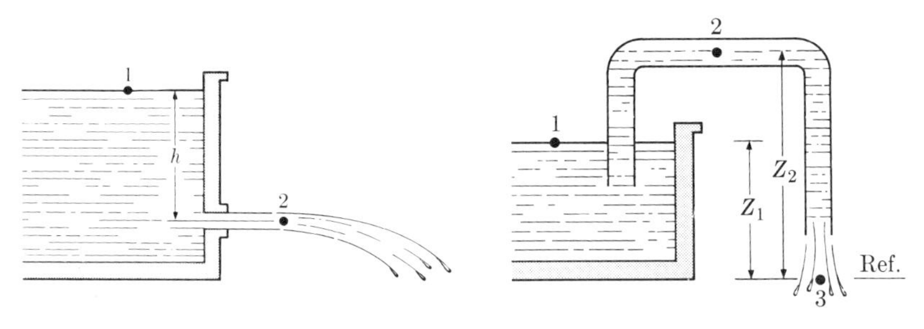

Fig. 6–1 Fig. 6–2

Torricelli's Equation

Figure 6–1 shows a reservoir, or fluid container, with an opening in its side. A jet of fluid flows from the opening. The surface of the fluid is at a height h above the centerline of the opening. We will consider a reference point (1) at the surface of the reservoir, and a second reference point (2) in the jet just beyond the opening—in the *free jet,* as it is called.

We can write Bernoulli's equation between these two points:

$$P_1/\gamma + v_1^2/2g + Z_1 = P_2/\gamma + v_2^2/2g + Z_2 + H_L.$$

By evaluating the parameters we may be able to solve for, or eliminate, some of the terms in the equation.

If we use an atmospheric reference, then $P_1 = 0$ psig, and hence $P_1/\gamma = 0$.

If the surface area of the reservoir is large relative to the area of the opening, and if h is constant, v_1 can be considered to be zero. Thus $v_1^2/2g = 0$.

When the level Z_2 is used as the reference, then $Z_1 = h$. Also, under these conditions, $Z_2 = 0$ ft or in.
In an ideal fluid condition, H_L is considered to be zero. Thus Bernoulli's equation reduces to

$$h = v_2^2/2g;$$

solving for the jet velocity v_2, we find

$$v_2 = \sqrt{2gh}, \qquad (6\text{–}1)$$

which is *Torricelli's theorem.*

Torricelli's theorem states that ideally the velocity of a free jet is equal to the square root of the product of two times the acceleration due to gravity times the head.

The Siphon

The familiar *siphon* is shown in Fig. 6–2. This is a continuous U-shaped tube with one end submerged beneath the surface of a liquid, the cross leg of the U above the level of the liquid surface, and the free end below it. Let us consider a siphon filled with fluid. When the fluid is allowed to run out of the free end, then the siphon will continue to discharge it. We can analyze the flow through a siphon, using Bernoulli's equation:

$$P_1/\gamma + v_1^2/2g + Z_1 = P_3/\gamma + v_3^2/2g + Z_3 + H_L.$$

Evaluating some of the parameters, as we did for Torricelli's equation, we get

$$P_1 = 0 \text{ psig} = P_3;$$

thus

$$\frac{P_1}{\gamma} = \frac{P_3}{\gamma} = 0 \text{ ft of head.}$$

As before, v_1, the surface velocity of the liquid, can be considered to be zero. Thus $v_1^2/2g = 0$. If Z_3 is used as the reference level, then $Z_1 = h$ and $Z_3 = 0$. For ideal flow, $H_L = 0$ ft.

When we substitute these values in Bernoulli's equation, we find that it reduces to $h = v_3^2/2g$; or solving for the velocity, we have

$$v_3 = \sqrt{2gh},$$

which is the same as the previously derived Torricelli equation. Note that we did not include in Bernoulli's equation any parameters from point 2, at the apex of the siphon. Whenever we write Bernoulli's equation, terms are included *only* from the two points being considered and not from any intermediate points!

Let us suppose next that we allow the head loss term, H_L, to assume a value; that is, $H_L \neq 0$. Bernoulli's equation then reduces to

$$h = \frac{v_3^2}{2g} + H_L,$$

and

$$v_3 = \sqrt{2g(h - H_L)}. \tag{6-2}$$

We see that the effect of the head loss is to reduce the *net head*, causing flow in the siphon; that is, the potential energy available to cause flow is reduced by the amount lost. The mechanism producing the loss will be considered later. Here we need be concerned only with the fact that it does exist.

Let's turn to the high point of the siphon, point 2. If we write Bernoulli's equation between points 1 and 2, we have

$$\frac{P_1}{\gamma} + \frac{v_1^2}{2g} + Z_1 = \frac{P_2}{\gamma} + \frac{v_2^2}{2g} + Z_2 + H_L.$$

Evaluating parameters as before and taking Z_1 as the reference, we reduce the equation to

$$P_2/\gamma = -v_2^2/2g - (Z_2 - Z_1) - H_L;$$

that is, a partial vacuum exists at point 2. This would be expected, since a reduced pressure would have to exist at point 2 in order for atmospheric pressure to cause the fluid to flow upward over the high point of the siphon. What is not so obvious is that the difference between atmospheric pressure and that at point 2 must also be great enough to provide for the velocity head and any losses.

If the distance $Z_2 - Z_1$ is made equal to 34 ft (ideally) and if $P_1 = 0$ psig (that is, if P_1 is atmospheric pressure), the siphon will not run. Atmospheric pressure will just support a column of water 34 ft high. If point 2 is 34 ft above point 1, there is no potential energy left to provide for $v_2^2/2g$ or H_L. We can see that the practical height limit is something less than 34 ft for water, depending upon the losses and the velocity of flow. The column height supported by atmospheric pressure head will, of course, be different for fluids with specific gravities other than unity.

EXAMPLE

Consider the siphon shown in Fig. 6–3. Using Bernoulli's equation, we will analyze the siphon in some detail; ideal flow (no losses) will be assumed. In order to evaluate the factors at any point in the siphon, it is necessary to know the velocity at that point as well as the pressure and elevation. To calculate the velocity, we must consider the free jet emerging from the end of the tube. We follow the step-by-step procedure outlined below.

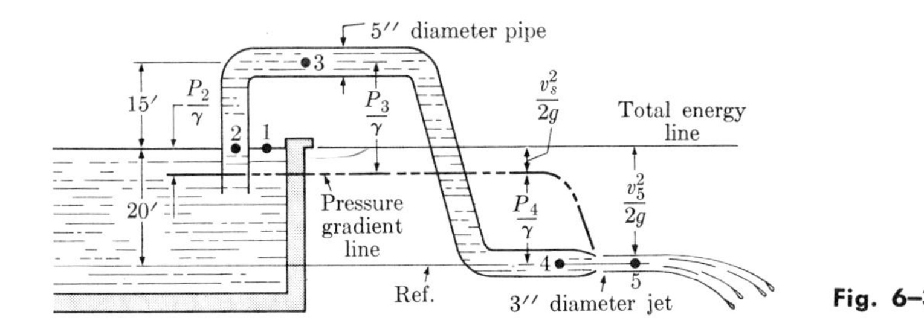

Fig. 6–3

Step 1. Write Bernoulli's equation between points 1 and 5:

$$P_1/\gamma + v_1^2/2g + Z_1 = P_5/\gamma + v_5^2/2g + Z_5,$$

$$0 + 0 + 20 \text{ ft} = 0 + v_5^2/64.4 + 0,$$

from which $v_5 = 35.82$ fps.

Step 2. The siphon tube is of constant cross section, 5 in. in diameter. The jet is 3 in. in diameter. Thus the velocity in the siphon tube can be calculated:

$$v_s = (\tfrac{3}{5})^2 v_5 = v_2 = v_3 = v_4 = 12.91 \text{ fps}.$$

Step 3. We can now proceed with a point-to-point analysis:

$$P_1/\gamma + v_1^2/2g + Z_1 = P_2/\gamma + v_2^2/2g + Z_2,$$

$$0 + 0 + 20 \text{ ft} = P_2/\gamma + (12.91)^2/64.4 + 20 \text{ ft};$$

$$P_2/\gamma = -167/64.4 = -2.582 \text{ ft}$$

or

$$P_2 = 0.433\, S_g h = 0.433 \times 1 \times -2.582 = -1.12 \text{ psi}.$$

This step calls attention to something which we should have felt intuitively—that a slight vacuum must exist just inside the entry to the siphon if the fluid is to flow. Flow always occurs from a point of higher potential (pressure) toward one of lower potential. We must lose some pressure between points 1 and 2, because we have gained some velocity head.

Step 4. Thus

$$P_1/\gamma + v_1^2/2g + Z_1 = P_3/\gamma + v_3^2/2g + Z_3,$$

$$0 + 0 + 20 \text{ ft} = P_3/\gamma + (12.91)^2/64.4 + 35 \text{ ft};$$

$$P_3/\gamma = -17.582 \text{ ft} \quad \text{or} \quad P_3 = -7.6 \text{ psi}.$$

Step 5. Consider now points 1 and 4:

$$P_1/\gamma + v_1^2/2g + Z_1 = P_4/\gamma + v_4^2/2g + Z_4,$$

$$0 + 0 + 20 \text{ ft} = P_4/\gamma + (12.91)^2/64.2 + 0 \text{ ft};$$

$$P_4/\gamma = +17.418 \text{ ft} \quad \text{or} \quad P_4 = 7.55 \text{ psi}.$$

Note that the pressure head at point 4 is reduced by the amount of velocity head in the siphon tube.

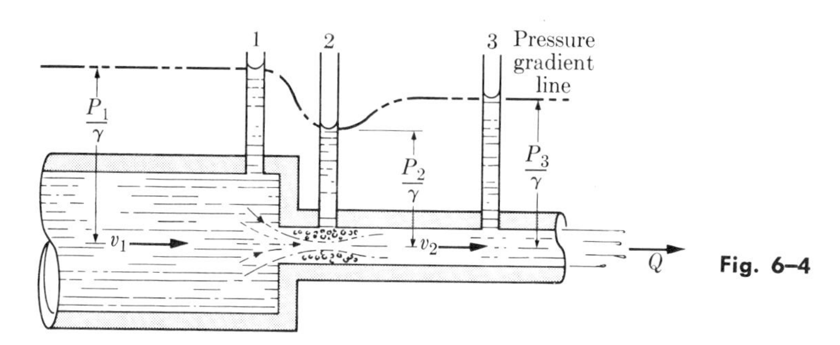

Fig. 6–4

Pressure Gradient

The term "gradient" means the rate of change of some variable with respect to another. In fluid mechanics we generally are interested in pressure gradient, which reflects the change of potential energy from point to point along a conductor. Figure 6–4 depicts our meaning of pressure gradient. We have a conductor of varying cross-sectional area. From principles of flow continuity we know that the velocity of flow will be greater in the smaller cross section than in the larger area. If we were to write Bernoulli's equation between sections 1 and 2, we would see that the pressure would drop as the velocity increased. Were piezometer tubes to be inserted at sections 1 and 2, the fluid would rise a distance equal to the pressure head, as shown in the Fig. 6–4. Let's move farther along the conductor to point 3.

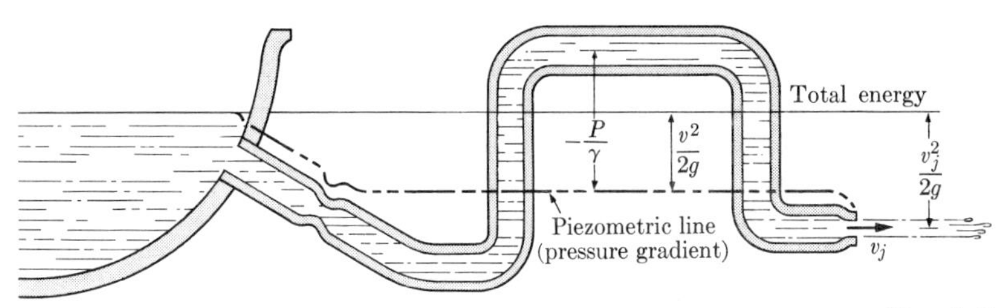

Fig. 6–5

A piezometer tube inserted here will also measure the pressure head. Under conditions of ideal flow, the height of the column of fluid in the piezometer tube at points 2 and 3 would be identical. Actually, we would experience some losses at point 2 due to the sudden change in cross section of the conductor. Thus $P_2/\gamma \neq P_3/\gamma$. If we now drew a line connecting the menisci of the fluid columns in the piezometer tubes, we would graphically construct the "pressure gradient or piezometric line" for this particular flow system. Figure 6–5 shows the relationships between the various heads and the gradient line in a system involving several basic elements we have discussed. Pressure gradients are not a major concern in fluid power systems because of the fact that operating pressures are quite high and pressure losses are relatively lower than for other types of fluid systems. Pressure gradient concepts are of importance in such civil engineering applications as water distribution systems, sewerage systems, etc.

Another interesting point is illustrated in Fig. 6–6, which summarizes some of the relationships between potential energy and kinetic energy; pressure head and velocity head; and elevation, pressure head, and velocity head. A column of fluid h units high represents a pressure head, $h = P/\gamma$, at the level of the centerline of the horizontal orifice. The same level is maintained at the opening of the vertical orifice. If we assume that the level of the fluid remains constant at h, then Torricelli's theorem tells us that the velocity of the jet issuing from the two orifices is $v = \sqrt{2gh}$ or $h = v^2/2g$. Thus the vertical free jet will rise to a height equal to h.

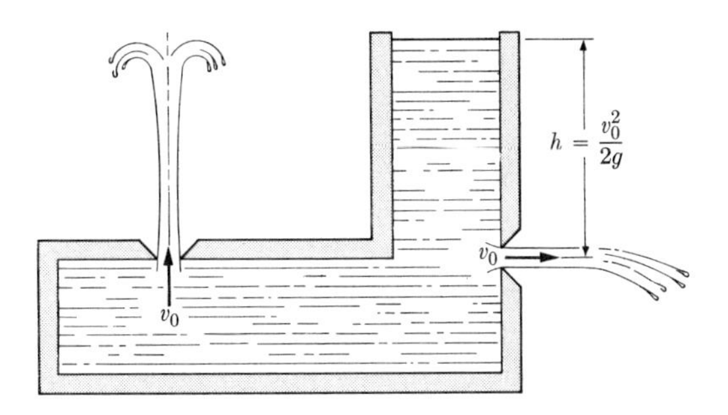

Fig. 6–6

IMPORTANT TERMS

Torricelli's theorem states that the velocity of a free jet is equal to the square root of the product of two times the acceleration due to gravity times the net head causing flow: $v = \sqrt{2gh}$.

Siphon is a continuous U-shaped conductor in which the discharge end is at a lower level than the fluid surface at the inlet end. Once flow has started, a siphon will continue to discharge due to the potential energy difference between the inlet and discharge end levels.

Free jet is a jet of fluid which is discharging into the atmosphere.

Pressure gradient is the rate of change of pressure along a conductor.

Piezometric line is a graphic plot of the pressure gradient.

PROBLEMS

6–1 A tank has in its side a hole 2 in. in diameter (see Fig. 6–7). The depth of water above the hole is 36 ft. Determine the velocity of the jet emerging from the hole.

6–2 What will be the flow rate Q_j in the jet of Problem 6–1?

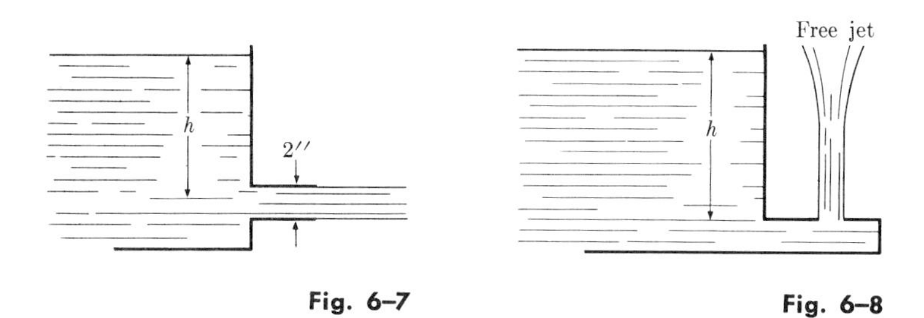

Fig. 6–7

Fig. 6–8

6–3 A large tank has a projection from the bottom, as shown in Fig. 6–8. An opening, or orifice, is located in the top surface. The depth of water-glycol hydraulic fluid ($S_g = 1.06$) above the orifice is 15 ft. If the orifice is 1 in. in diameter, what is the velocity of the jet issuing from it? What is the theoretical flow rate? How high will the jet rise?

6–4 If the head loss across the orifice of Problem 6–3 is $H_L = 2$ ft, what will be the velocity of the jet? How high will it rise?

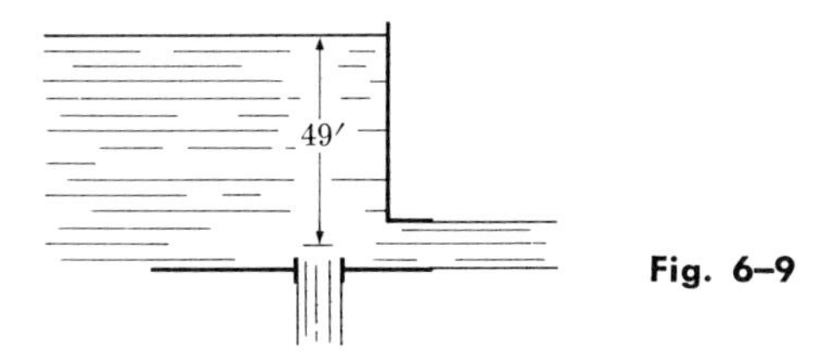

Fig. 6–9

6–5 A reservoir has a horizontal opening and a downward vertical opening, as shown in Fig. 6–9. Calculate the jet velocity for both openings.

6–6 A standpipe filled with fluid has openings at the quarter lengths, as shown in Fig. 6–10. Assuming that the fluid level remains constant, calculate the points at which the jets of fluid will strike the ground at the level of the base of the standpipe.

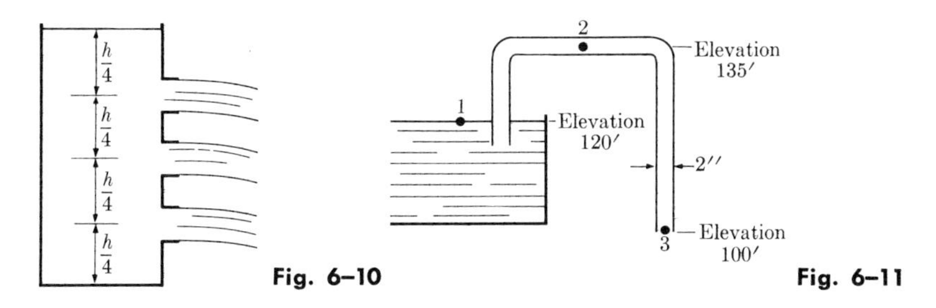

Fig. 6–10

Fig. 6–11

6–7 Calculate the flow rate of the fluid (water) issuing from the siphon shown in Fig. 6–11.

6–8 In Fig. 6–11, if $H_{L_{1-2}} = 2$ ft and $H_{L_{2-3}} = 1.5$ ft, what would be the flow rate?

6–9 Calculate the absolute pressure at point 2 in the siphon in Fig. 6–11.

6–10 Figure 6–12 illustrates a siphon system in which $H_{L_{1-2}} = 1.5$ ft, $H_{L_{2-3}} = 0.5$ ft, $H_{L_{3-4}} = 3$ ft. Calculate the flow rate of discharge of water. Calculate flow velocities at points 2, 3, and 4. What are the pressures at points 2 and 3?

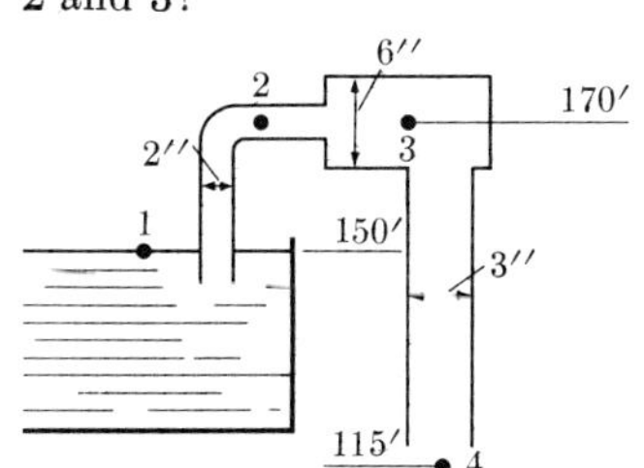

Fig. 6–12

6–11 Suppose that we put a cover over the reservoir in Fig. 6–11, and let the siphon tube pass through it. If we then pressurize the space in the reservoir above the fluid surface to 5 psi, what will be the flow rate in the siphon?

6-12 A sealed chamber is shown in Fig. 6-13. A partition divides the chamber into two sections, and a siphon passes through the partition. Pressures and elevations are shown in the sketch. Calculate the flow rate through the siphon. What is the pressure at point 2? Oil is flowing from the left to the right side of the chamber. At what height would the flow of oil stop?

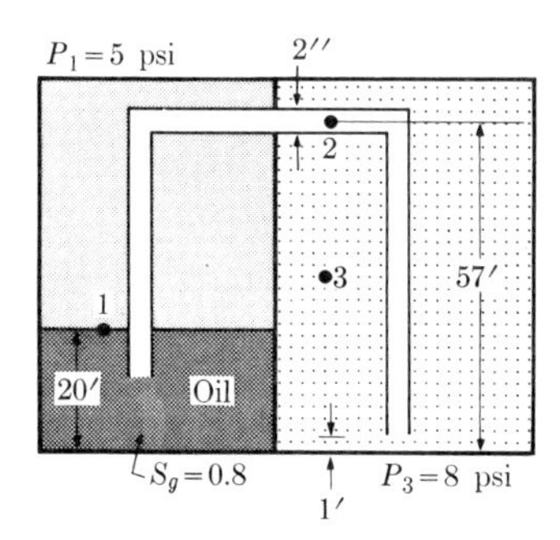

Fig. 6-13

6-13 Derive the algebraic relationship involving pressure on the surface of the fluid and the elevation difference $Z_2 - Z_3$, which causes the pressure at point 2 to become absolute zero. Use a typical siphon configuration as in Fig. 6-11.

CHAPTER 7

More applications of Bernoulli's equation

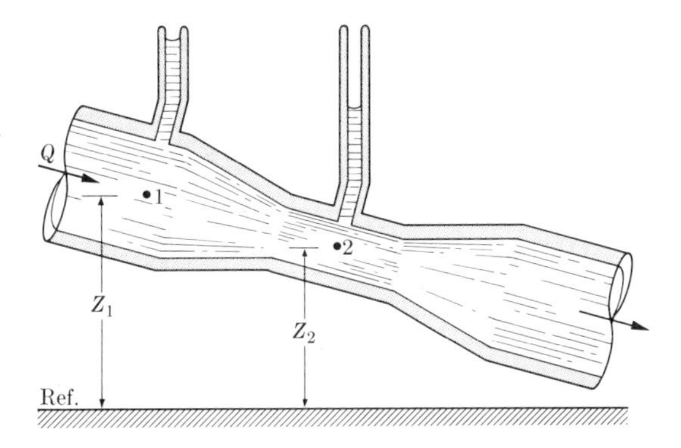

Fig. 7–1

Bernoulli's equation is one of the basic relationships of fluid mechanics and can be used to solve problems involving flow systems other than the two simple examples treated in the previous chapter.

The Venturi

In earlier chapters we have discussed continuity of flow and conductors of varying cross sectional area. Now let us examine the special case illustrated in Fig. 7–1, which shows a circular conductor with an area of reduced cross section and a gradual transition from the large upstream area to the reduced area. The transition from the reduced area to the enlarged downstream section is also gradual. Such a device is called a *venturi*. Application of Bernoulli's equation to the analysis of flow through the venturi will reveal some of the interesting characteristics of this device. Assuming that point 1 is in the large upstream area, and point 2 is in the reduced central area, called the *throat*, we can write

$$P_1/\gamma + v_1^2/2g + Z_1 = P_2/\gamma + v_2^2/2g + Z_2 + H_L.$$

Note that if we group all the kinetic energy terms on one side of the equation and all the potential energy terms on the other, we have

$$\frac{v_2^2 - v_1^2}{2g} = (P_1/\gamma + Z_1) - (P_2/\gamma + Z_2 + H_L). \tag{7-1}$$

This relationship shows that in the flowing system

change in kinetic energy = change in potential energy:

$$\Delta \text{ KE} = \Delta \text{ PE}.$$

Thus the increase in kinetic energy at the throat is equal to the decrease in potential energy at the throat. From this "Venturi principle" we would expect to find a reduction in pressure at the throat; that is, $p_1 > p_2$.

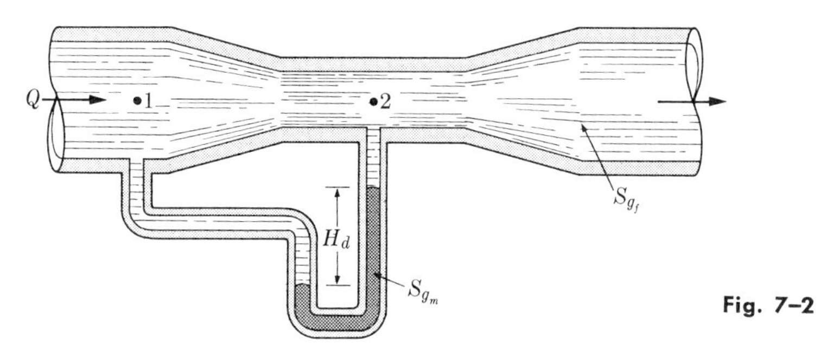

Fig. 7–2

Just how great this reduction in pressure will be can be determined by solving Bernoulli's equation for any given system. Let us consider Fig. 7–2, which shows a venturi with a manometer connected between the upstream section and the throat. This device is called a *venturi meter*. When properly constructed and calibrated, a venturi meter can be used to meter (measure) flow rate and/or to control flow processes. Since the venturi meter of Fig. 7–2 is horizontal (as is usually the case), $Z_1 = Z_2$. For ideal flow, $H_L = 0$. Then Bernoulli's equation reduces to

$$P_1/\gamma + v_1^2/2g = P_2/\gamma + v_2^2/2g,$$

or

$$\frac{v_2^2 - v_1^2}{2g} = \frac{P_1 - P_2}{\gamma} = H_d. \tag{7-2}$$

From the principle of flow continuity, $A_1v_1 = A_2v_2$, we can solve for $v_2 = (A_1/A_2)v_1$. Substituting this expression for v_2 in Eq. (7–2), we get

$$\frac{(A_1/A_2)^2v_1^2 - v_1^2}{2g} = \frac{P_1 - P_2}{\gamma}.$$

Solving for v_1, we obtain

$$v_1 = \sqrt{\frac{2g}{(A_1/A_2)^2 - 1}\,\frac{P_1 - P_2}{\gamma}}. \tag{7–3}$$

For a given venturi, A_1 = const and A_2 = const. Thus we could set

$$\sqrt{\frac{1}{(A_1/A_2)^2 - 1}} = K,$$

and Eq. (7–3) would reduce to

$$v_1 = K\sqrt{2g\,\frac{(P_1 - P_2)}{\gamma}}. \tag{7–4}$$

Note that this expression has the general form of the Torricelli equation.

From continuity of flow, we have $Q = A_1v_1$. Substituting v_1 by Eq. (7–4) gives

$$Q = A_1K\sqrt{2g\,\frac{(P_1 - P_2)}{\gamma}}. \tag{7–5}$$

The manometer head difference H_d, when reduced to consistent units, is equal to the pressure head difference. Thus $H_d = (p_1 - p_2)/\gamma$. And substituting in (7–5), we have

$$Q = A_1K_1\sqrt{2gH_d}, \qquad \text{where } K_1 = K\sqrt{(S_{g_m} - S_{g_f})/S_{g_f}}.$$

It can be seen that the manometer head difference, H_d, can be calibrated to read in terms of units of flow rate, such as gpm, ft^3/min, in^3/min, etc. The venturi, as such, finds applications as a flow rate meter or flow control device.

EXAMPLE

Using a typical venturi, such as is shown in Fig. 7–2, assume that the diameter of the conductor is 2 in. at point 1 and 1 in. at point 2. If the manometer is filled with mercury and the head difference is 21.43 in., what is the approximate flow rate of water through the venturi? What is the flow velocity at point 1 and at point 2 in the conductor?

Solution.

Step 1. The head difference is given in inches of mercury. Since the flowing fluid is water, the H_d must be converted to an equivalent head of water:

$$H_{d_{\text{water}}} = S_g \times H_{d_{\text{mercury}}} = 13.6 \times 21.43 \text{ in.} = 292 \text{ in. of water.}$$

Step 2. From Eq. (7–2) we see that the pressure head difference is equal to the velocity head difference and that these are both equal to H_d:

$$\frac{v_2^2 - v_1^2}{2g} = \frac{P_1 - P_2}{\gamma} = H_d.$$

Then

$$\frac{v_2^2 - v_1^2}{2g} = 292 \text{ in.}$$

Step 3. Since neither v_2 nor v_1 is known, we must eliminate one of the variables:

$$v_2 = \frac{A_1}{A_2} v_1 \qquad \text{and} \qquad \frac{A_1}{A_2} = \frac{(D_1)^2}{(D_2)^2}.$$

Then

$$\frac{(D_1/D_2)^4 v_1^2 - v_1^2}{2g} = 292;$$

$$\frac{[(2/1)^4 - 1]v_1^2}{2 \times 386} = 292,$$

from which

$$v_1 = \sqrt{\frac{292 \times 772}{16 - 1}} = \sqrt{15{,}050} = 122.8 \text{ ips.}$$

Step 4. Calculate v_2:

$$v_2 = (D_1/D_2)^2 v_1 = 4 \times 122.8 = 490 \text{ ips.}$$

Step 5. Calculate the flow rate from the principle of flow continuity:

$$Q = A_1 v_1 = 3.1416 \text{ in}^2 \times 122.8 \text{ ips} = 385 \text{ in}^3/\text{sec}$$

Conversion to gpm gives

$$1 \text{ gal} = 231 \text{ in}^3,$$

$$Q = 385/231 \times 60 = 100 \text{ gpm.}$$

Note that in the development of the preceding example, we have treated only ideal flow. This approach will give results which are theoretically correct but cannot be matched by actual performance. Practical and theoretical performances do not agree because of the existence of flow losses H_L and deviations from exact dimensions in the actual venturi meter due to manufacturing tolerances. To take account of these unavoidable deviations in actual instruments, each meter is calibrated after it has been completed.

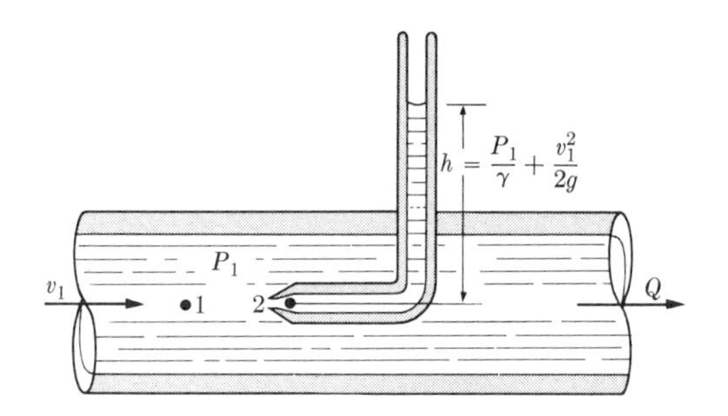

Fig. 7–3

The Pitot Tube

The device shown in Fig. 7–3 is called a *Pitot tube.* Its function is to measure the velocity of fluid flow. As shown in the figure, the Pitot tube is installed in a closed conductor. It is a streamlined tube, with its axis parallel to the direction of fluid flow, and with an opening in the tip on which the fluid impinges. This section of the tube is connected to a head-sensing device outside the conductor; in our illustration the device is a piezometer tube. We can analyze the flow through a Pitot tube by writing Bernoulli's equation between point 1, which is in the conductor upstream from the tube, and point 2, which is just inside the opening in the tip:

$$P_1/\gamma + v_1^2/2g + Z_1 = P_2/\gamma + v_2^2/2g + Z_2.$$

Since points 1 and 2 are at the same level, we find $Z_1 = Z_2$. Also note that there is no flow inside the Pitot tube; if there were, the tube would overflow. Hence $v_2 = 0$. Bernoulli's equation reduces to

$$P_2/\gamma = P_1/\gamma + v_1^2/2g.$$

Thus the height of the column of fluid in the piezometer tube (the head) is equal to the sum of the pressure head and the velocity head inside the conductor.

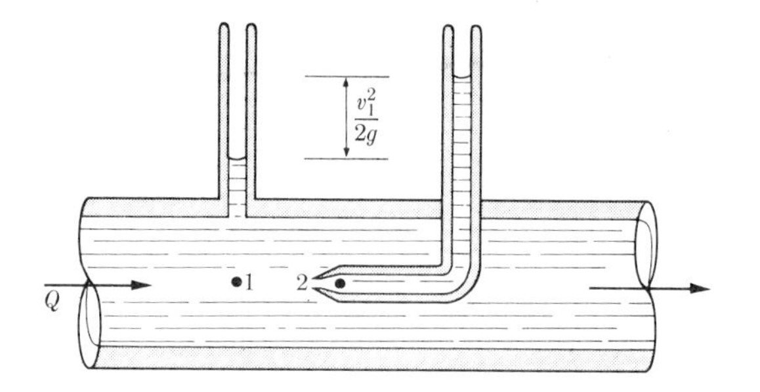

Fig. 7–4

In order to determine the velocity, these two components of the head must be separated. One way of doing this is to insert a second piezometer tube into the wall of the conductor above point 1, as shown in Fig. 7–4. This second tube indicates the pressure head at point 1, P_1/γ. If the two piezometer tubes are now connected across a manometer, the remaining head difference will be the velocity head, since there is a component of pressure head in each tube. Note that if we solve the difference relationship for velocity between velocity head and manometer head, we get the Torricellian equation for velocity, $v = \sqrt{2gh_d}$.

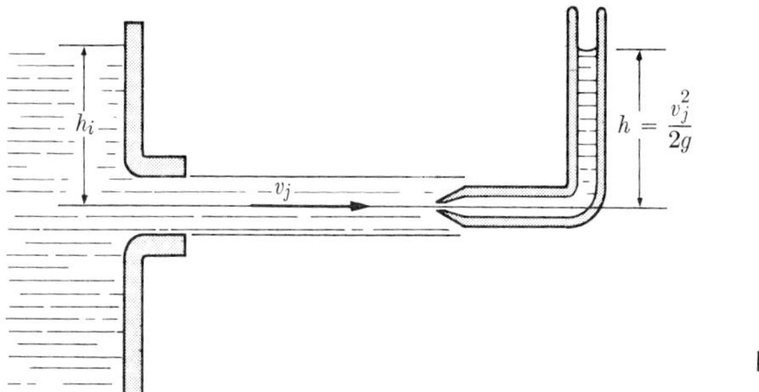

Fig. 7–5

If a Pitot tube is introduced into a free jet which, uninhibited by a conductor, emerges into the atmosphere (as shown in Fig. 7–5), the pressure head in the jet is zero. The Pitot tube thus measures directly the jet velocity head.

One of the best known applications of the Pitot tube is in the measurement of air speed of airplanes. Pitot tubes are also used to measure air flow velocity in ducts, discharge velocity from fans, etc.

EXAMPLE

Assume that there is a free jet (Fig. 7–5) and that a Pitot tube is placed in the jet. The fluid rises to a height of 100 in. in the piezometer tube. If the fluid is water, what is the velocity of the jet? If the diameter of the jet is 2 in., what is the flow rate in gpm?

Solution.

We proceed in two steps.

Step 1. Since the entire head measured by the Pitot tube in a free jet is velocity head, the Torricelli equation applies:

$$v = \sqrt{2gh} = \sqrt{2 \times 386 \times 100}$$
$$= \sqrt{77{,}200} = 278 \text{ ips} \quad \text{or} \quad 23.2 \text{ fps.}$$

Step 2. From continuity of flow, $Q = Av$:

$$Q = \frac{\pi d^2}{4} v = 3.1416 \times 278 = 873 \text{ in}^3/\text{sec}$$

$$Q = \tfrac{873}{231} \times 60 = 227 \text{ gpm.}$$

With the establishment of the idea of a Pitot tube, we can proceed to Fig. 7–6, which is intended to summarize a number of the concepts discussed and centers around Bernoulli's equation and conservation of energy. The figure represents a conductor of nonuniform cross-sectional area. At three random points along the conductor, points 1, 2, and 3, we want to sample the condition of the various components of total energy. It is fairly simple to measure the elevations of the three points: Z_1, Z_2, and Z_3. These represent potential energy (per pound of fluid flowing) due to elevation above some datum level. By means of piezometer tubes inserted at the

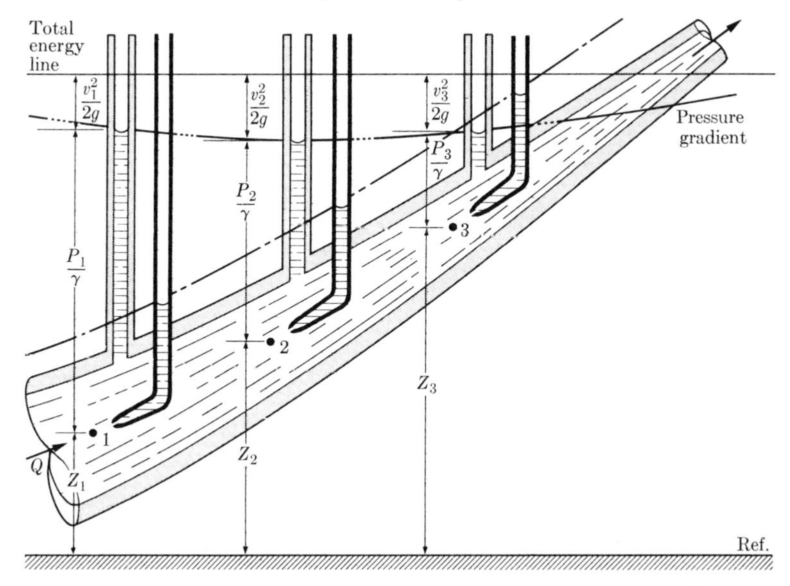

Fig. 7–6

three reference points we can determine the pressure heads P_1/γ, P_2/γ, and P_3/γ. These, you will remember, represent potential energy (per pound of fluid flowing) due to pressure.

To illustrate our next point, we must remember that we indicated that a Pitot tube inserted into a conductor measures *both* pressure and velocity heads. For our purposes, we will assume that the Pitot tubes in Fig. 7-6 measure only velocity head. By doing this, we are able to visualize how the velocity head varies along the conductor as the cross section changes. Hence velocity heads $v_1^2/2g$, $v_2^2/2g$, and $v_3^2/2g$ represent the kinetic energy (per pound of fluid flowing) due to the flow velocity at each point. (Figure 7-4 shows what we would actually have to do to see the velocity head.)

We have now graphically illustrated the three major components of total energy in the flowing stream. Pressure energy provides the potential needed to cause flow along the conductor. It must supply the energy to provide for any changes in elevation and kinetic energy due to changes in flow velocity. Thus from point 1 to point 2, the elevation increases from Z_1 to Z_2, and the velocity increases from v_1 to v_2 because of the reduction in cross section. The pressure head must *decrease* from P_1/γ to P_2/γ to provide the energy for increases in elevation and kinetic energy. The same analysis applies in going from point 2 to point 3. Assuming ideal flow, we find that the sum of the heads at each point is constant; the heads will always equal the total energy level.

IMPORTANT TERMS

Venturi is the name applied to the conductor configuration in which there is a gradual transition from the full diameter to a reduced diameter section, called the throat, and subsequently a gradual return to the full diameter.

Venturi principle states that the increase in kinetic energy in the throat, due to increased velocity, is equal to the decrease in potential energy (pressure) in the throat.

Venturi meter is a venturi built and calibrated to measure the flow rate of a fluid.

Pitot tube is a device which measures flow velocity by conversion of kinetic energy of flow in the conductor to potential energy (head) in a connected piezometer tube or similar device.

PROBLEMS

7–1 A venturi consists of a pipe section 3 in. in diameter—and a throat section of $1\frac{1}{2}$ in. in diameter. The upstream pressure is 50 psi, and the flow rate (water) is 150 gpm. Calculate the pressure in the throat area.

7–2 A conductor is 2 in. in diameter. The pressure upstream from a reduced section in the conductor is 80 psi, and the pressure at the reduced section is 60 psi. The flow rate of phosphate ester hydraulic fluid ($S_g = 1.1$) is 80 gpm; calculate the diameter of the reduced section.

7–3 A venturi with large-section diameter of 2 in. and throat diameter of 1 in. passes 75 gpm of water-in-oil hydraulic fluid ($S_g = 0.9$). Upstream pressure is 150 psi. Assume that a piezometer tube is inserted in both sections of the venturi. How high will the fluid rise in each piezometer tube?

7–4 If the piezometer tubes of Problem 7–3 were connected across the legs of a mercury manometer, what would be the differential head displayed by the manometer?

7–5 A venturi is 3 in. in diameter at the large section; pressure is 100 psi at the 3-in. section and 75 psi at the throat; 90 gpm of mineral base hydraulic fluid ($S_g = 0.89$) passes through the venturi. Calculate the diameter of the throat.

7–6 Since the pressure drop in a venturi between the full-size section and the throat is a function of the flow rate through the venturi, the drop can be used to indicate flow rate. Figure 7–2 illustrates a typical configuration for a venturi meter. Assume that the diameter is 2 in. at point 1 and 1 in. at point 2. Using water as the fluid, calculate the head differentials, H_d, for a mercury manometer over a range of flow rates. Plot a curve of head differential H_d vs. flow rate. Of what use might such a curve be in an actual application of a venturi meter?

7–7 Suppose that in Problem 7–6 the fluid had been glycerine instead of water. What would be the slope of the H_d-vs.-Q curve? What is the difference in percent between this curve and that of Problem 7–6? What does this suggest as to the accuracy of metering flow rates of fluids other than the one for which the venturi meter is calibrated?

7–8 What would be the effect of head losses in the venturi on the accuracy of flow measurement? Since head losses are difficult to predict, how would one go about calibrating an actual venturi meter?

7–9 Figure 7–3 illustrates a typical Pitot tube. If the pipe is 4 in. in diameter, the pressure is 100 psi, and the flow rate is 250 gpm, how high in the Pitot tube will the fluid rise? Assume that the fluid is sea water.

7–10 A typical Pitot tube is inserted in a 2 in. pipe. The pressure is 20 psi. Water rises to a height of 50 ft in the vertical leg of the tube. What is the flow rate in the pipe?

7–11 The device shown in Fig. 7–7 is called an *aspirator*. The flow rate Q in the main conductor causes a reduction in pressure at point 2 in the secondary

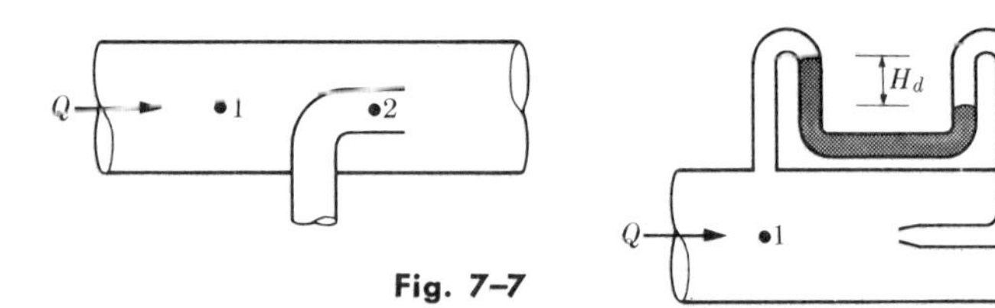

Fig. 7–7

Fig. 7–8

conductor. This reduction in pressure produces a secondary flow, Q_a. Thus fluid can be "pumped" by means of an aspirator. The pipe diameter is 3 in. at point 1 and 1 in. at point 2. Calculate the pressure at point 2 when $Q = 120$ gpm, and $P_1 = 100$ psi.

7–12 In Fig. 7–8, the Pitot tube and piezometer tube installed in a pipe are connected across a mercury manometer. What quantity will the head difference H_d, represent?

7–13 In Fig. 7–8, the pipe diameter is 3 in. The pressure at point 1 is 25 psi, and the flow rate is 200 gpm. What will be the head difference, H_d, if the gage fluid is mercury and if the working fluid is a silicone fluid ($S_g = 1.03$)?

7–14 Given an installation similar to that shown in Fig. 7–8: the pipe is 2 in. in diameter, the fluid is water, the pressure at point 1 is 40 psi, $H_d = 3.66$ in. Hg. Calculate the flow rate in the pipe.

7–15 As in Fig. 7–8, a Pitot tube is installed in a pipe to measure flow rate of water as a function of flow velocity. In other words, it is used as a flow meter. Using the parameters of Problem 7–12, calculate the head difference in the differential manometer for a range of flow rates. Plot a curve H_d vs. Q. What is the significance of this curve?

7–16 A Pitot tube is installed in a free jet of water, as shown in Fig. 7–5. If the water rises to a height of 3 ft, what is the velocity of the jet?

7–17 If the jet flow velocity in the Pitot tube of Problem 7–16 is 100 fps, how high will the water rise in the vertical leg?

7–18 A free jet emerges from a nozzle 2 in. in diameter. The flow rate through the jet is 100 gpm. How high will the fluid rise in the vertical leg of a Pitot tube inserted into the jet?

7–19 A Pitot tube is suspended from a boat traveling at 30 mph so that the tube passes through the water at a shallow depth. How high will the water rise in the vertical leg of the Pitot tube? How could this effect be used to indicate the speed of the boat?

7–20 Pitot tubes are used to measure the air speed of airplanes. Explain how the tubes might be applied to measure air speed.

7–21 A manufacturer of an automatic transmission for automobiles uses a Pitot tube positioned on the rim of a flywheel to sense rotational speed. Develop a simple expression for the relationship between flywheel diameter, output speed, velocity head, and pressure head in the Pitot tube.

CHAPTER 8

Orifices

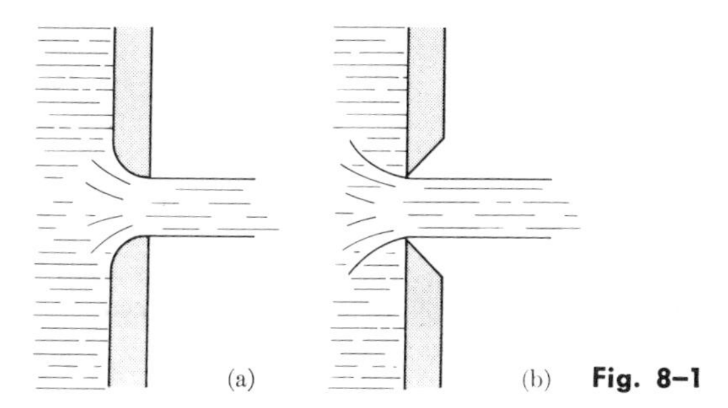

Fig. 8–1

What Is An Orifice?

In general terms, we might think of an orifice as a hole through which fluid flows. Although we tend to think of orifices as being round, they can also be square or triangular. As we shall demonstrate later, flow passages somewhat more complex than simple holes in relatively thin plates can sometimes be characterized as orifices, and their behavior can be described in the same general way.

Initially we shall consider round orifices in thin plates. By a thin plate we mean one in which the thickness of the material is small relative to the diameter of the opening. Figure 8–1 illustrates two conditions under which such an orifice might exist. Figure 8–1(a) shows a so-called *round-edged* orifice; Fig. 8–1(b) illustrates a *sharp-edged* orifice.

Sharp-Edged Orifices

These orifices have some special characteristics which make them useful as flow-measuring and control devices. Thus, for example, a true sharp-edged orifice is viscosity insensitive: that is, its performance is not affected by changes in temperature of the fluid. Hence the flow characteristics of an orifice can be described in terms of dimensionless coefficients.

In practice, it is difficult to obtain a true sharp-edged orifice. When orifices are used as flow-metering devices, each one must be calibrated to determine its actual performance under operating conditions.

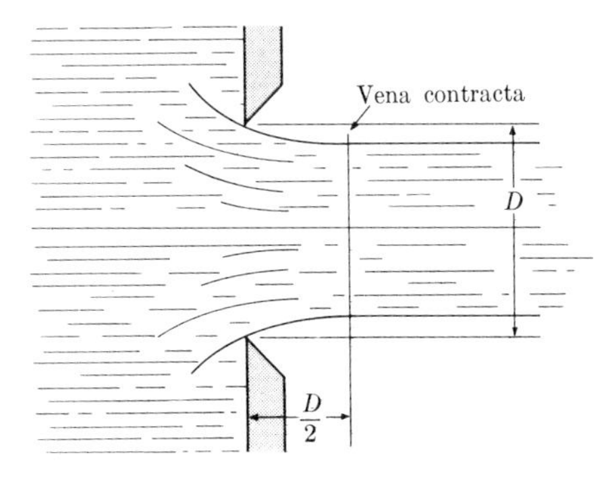

Fig. 8–2

Figure 8–2 shows a typical sharp-edged orifice under flow conditions. The curved lines in the illustration represent streamlines of fluid flowing through the orifice. Note that the fluid flowing adjacent to or near the surface of the plate must undergo a much greater change of direction than that flowing toward the center of the orifice. Because of its momentum, the flowing fluid cannot make the sharp turn required for it to exit from the orifice at right angles to the plate. Thus the jet issuing from the orifice contracts. At some point downstream from the plane of the orifice opening, the jet achieves a constant cross section and does not contract any more. This point, or section, is called the *vena contracta*. It has been demonstrated empirically that the vena contracta occurs at a section $D/2$ downstream from the plane of the opening (D is the diameter of the orifice opening).

We can define a *coefficient of contraction*, C_c:

$$C_c = A_j/A_o$$

where A_j = cross-sectional area of the jet at the vena contracta, and
A_o = cross-sectional area of the actual orifice opening.

If we define a quantity v_j as the actual mean velocity in the jet at the vena contracta, we can then define a *velocity coefficient* as

$$C_v = v_j/v_t$$

where v_j = actual velocity in the jet, and
v_t = theoretical or Toricellian velocity $= \sqrt{2gh}$.

From continuity of flow principles, we can write an expression for the flow rate through the orifice:

$$Q_o = A_j v_j = C_c A_o \cdot C_v \sqrt{2gh} = C_c C_v \cdot A_o \sqrt{2gh}.$$

If we define a *discharge coefficient,* $C_d = C_c \cdot C_v$, we can reduce the expression above to the general form for flow through a sharp-edged orifice:

$$Q = C_d A_o \sqrt{2gh}. \tag{8-1}$$

Equation (8–1) tells us that the flow rate through an orifice is equal to the product of some empirically determined discharge coefficient, the cross-sectional area of the orifice, and the square root of the product of twice the acceleration due to gravity and the net head causing flow. The important point to note here is that the flow rate is a function of the square root of the head—generally, the pressure head causing flow. This makes the flow through an orifice nonlinear, a fact which causes some concern in applications of this device to fluid-power control devices. A further complication is the virtual impossibility of making a flow-control orifice in a practical control device a true sharp-edged orifice. Thus these devices are sensitive to changes in temperature and their effect on fluid properties.

Approximate values, empirically determined, for the sharp-edged orifice coefficients are

$$C_c = 0.60 \text{ to } 0.62, \qquad C_v = 0.98, \qquad C_d = 0.60.$$

EXAMPLE

In the reservoir of Fig. 6–1, let $h = 100$ ft. What will be the discharge flow rate issuing from a sharp-edged orifice 2 in. in diameter in the side of the reservoir?

Solution.

We proceed in steps as follows.

Step 1. Write the equation for discharge from an orifice

$$Q = C_d A_o \sqrt{2gh}.$$

Step 2. Substitute the given numerical values into above equation:

$$Q = 0.60 \times A_o \sqrt{2 \times 32.2 \times 100}.$$

Step 3. Keeping units consistent, solve for unknown parameters, in this case the orifice area:

$$A_o = \pi d^2/4 = (\pi 2^2/4)(1/144) = 3.1416/144 = 0.0218 \text{ ft}^2.$$

Step 4. $Q = 0.60 \times 0.0218 \times \sqrt{64.4 \times 100} = 1.048 \text{ ft}^3/\text{sec}.$

Step 5. Check dimensionally:

$$\begin{aligned} \text{ft}^3/\text{sec} &= \text{dimensionless number} \times \text{ft}^2 \times \sqrt{(\text{ft}/\text{sec}^2) \times \text{ft}} \\ &= \text{ft}^2 \times (\text{ft}/\text{sec}) = \text{ft}^3/\text{sec}. \end{aligned}$$

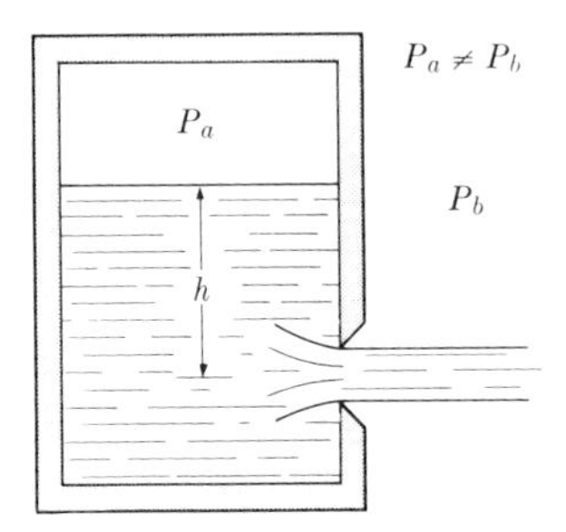

Fig. 8–3

Concept of Net Head

In the preceding discussion we have considered only the simple case of a free jet issuing from an orifice with atmospheric pressure on the surface of the fluid in the reservoir. Suppose that we have a case such as is shown in Fig. 8–3, where there is a pressure P_a on the surface of the liquid and the jet is discharging into an atmosphere at pressure P_b, such that $P_a \neq P_b$.

Writing Bernoulli's equation and solving for the velocity of the jet yields an expression of the form

$$v_t = \sqrt{2g[h + (P_a - P_b)/\gamma]}.$$

This is still in Torricellian form (Eq. 6–1), but the *net head* is now expressed in two terms instead of one. The first is due to the height of the column of fluid above the orifice, h, and the second is due to the difference in pressure heads,

$$(P_a - P_b)/\gamma.$$

Concept of Approach Velocity

Let us now consider a case in which the orifice at the end of a pipe (see Fig. 8–4), is such that the diameter of the pipe is not great compared to the diameter of the orifice. In such cases we cannot neglect the approach velocity of the fluid flowing through the system as we did when considering flow from an orifice in the wall of a reservoir. Now when we write Bernoulli's equation and solve it for jet velocity, we get

$$v_t = \sqrt{2g(P_1/\gamma + v_1^2/2g)},$$

which is Torricellian in form. The net head now includes a term for the pressure head, P_1/γ, and a term for the approach velocity head, $v_1^2/2g$.

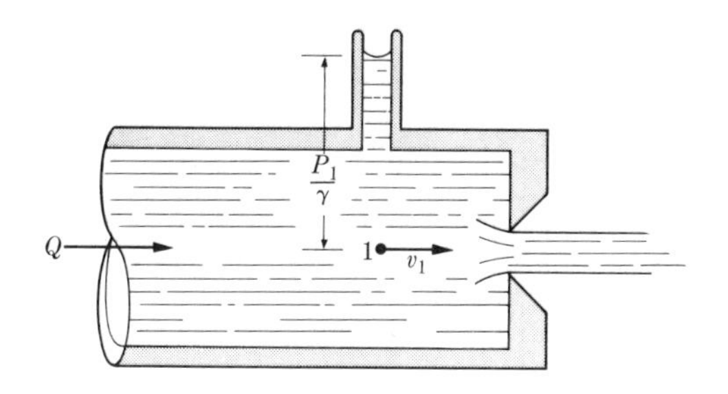

Fig. 8–4

We can write the discharge rate equation for Fig. 8–4 as follows:

$$Q = C_d A_o \sqrt{2g(P_1/\gamma + v_1^2/2g)}. \tag{8-2}$$

Equation (8–2) contains a term which is a function of the flow rate. We can eliminate this term and derive an expression entirely in system parameters as follows: From $Q = A_1 v_1$, we have

$$v_1 = Q/A_1 = \frac{Q}{\pi d_1^2/4}.$$

Substituting in (8–2), yields $Q = C_d A_o \sqrt{2g[P_1/\gamma + Q^2/(\pi d_1^2/4)^2/2g]}$, but $A_o = \pi d_o^2/4$. Substituting and squaring the above equation, we obtain

$$Q^2 = C_d^2 \left(\frac{\pi}{4} D_o^2\right)^2 2g \left\{\frac{P_1}{\gamma} + \frac{Q^2/[(\pi/4)D_1^2]^2}{2g}\right\},$$

$$Q^2 = 2g\left(\frac{P_1}{\gamma}\right) C_d^2 \left(\frac{\pi D_o^2}{4}\right)^2 + 2g\left(\frac{\pi D_o^2}{4}\right)^2 C_d^2 \left(\frac{Q^2}{2g[\pi D_1^2/4]^2}\right),$$

from which

$$Q^2 = 2g\frac{P_1}{\gamma} C_d^2 A_o^2 + C_d^2 Q^2 \left(\frac{D_o}{D_1}\right)^4,$$

$$Q^2\left[1 - C_d^2\left(\frac{D_o}{D_1}\right)^4\right] = 2g C_d^2 A_o^2 \frac{P_1}{\gamma}.$$

Solving for Q, we have

$$Q = C_d A_o \sqrt{2g(P_1/\gamma)} \left[1 - C_d^2 \left(\frac{D_o}{D_1}\right)^4\right]^{-1/2}. \tag{8–3}$$

Note that Eq. (8–2) is still in Torricellian form but has an additional factor. We might think of this as a correction factor to accommodate the approach velocity head.

To get this factor in a more usable form we can expand it by means of the binomial theorem, as follows:

$$\left[1 - C_d^2 \left(\frac{D_o}{D_1}\right)^4\right]^n = 1^n + n \cdot 1^{(n-1)} \cdot C_d^2 \left(\frac{D_o}{D_1}\right)^4 + \frac{n(n-1)}{2!} \cdot 1^{(n-2)} \cdot \left[C_d^2 \left(\frac{D_o}{D_1}\right)^4\right]^2 + \cdots$$

$$= 1 + \frac{1}{2} \cdot C_d^2 \left(\frac{D_o}{D_1}\right)^4 + \cdots.$$

We can neglect the additional terms beyond the second to simplify the expression without introducing undue inaccuracy. The final equation for the flow through an orifice where approach velocity is significant is shown in the following equation:

$$Q = C_d A_o \sqrt{2g(P_1/\gamma)} \left[1 + \frac{1}{2} \cdot C_d^2 \left(\frac{D_o}{D_1}\right)^4\right]. \tag{8–4}$$

Orifice in a Pipe

In Fig. 8–4 we dealt with the case of a free jet issuing from the orifice at the end of a conductor. Figure 8–5 illustrates an orifice placed within the pipe itself. This is the usual application for an orifice used as a metering or control device. Since we are still dealing with ideal flow, we can approach the orifice* from the point of view of Bernoulli's equation. In the case of the free jet, the pressure head causing flow was P_1/γ; here, however, flow is determined by the differential pressure head, $(P_1 - P_2)/\gamma$. Note that as the jet issues from the orifice, it causes turbulence beyond the orifice plate, and further reduces the pressure at point 2. Equation (8–4) becomes

$$Q = C_d A_o \sqrt{2g(P_1 - P_2)/\gamma}\, [1 + \tfrac{1}{2} C_d^2 (D_o/D_1)^4]. \tag{8–5}$$

Since the orifice induces a pressure differential as a function of flow, it can be used to measure flow in much the same manner as the venturi meter

* As a matter of fact, Fig. 8–5 represents an orifice involving a *submerged jet*, which is a topic beyond the scope of this text.

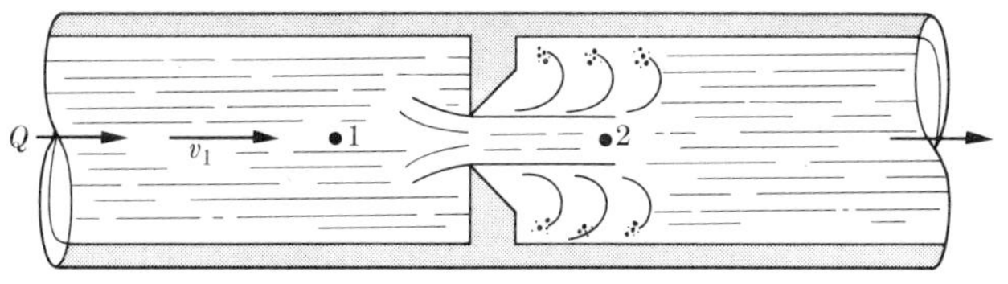

Fig. 8–5

discussed previously. Pressure taps can be placed in the conductor upstream and downstream from the orifice, and the pressure heads can be impressed across a differential manometer measured at these two points. The head difference can be calibrated to read in flow rate terms, such as gal/min, ft^3/sec, etc. See Fig. 8–6.

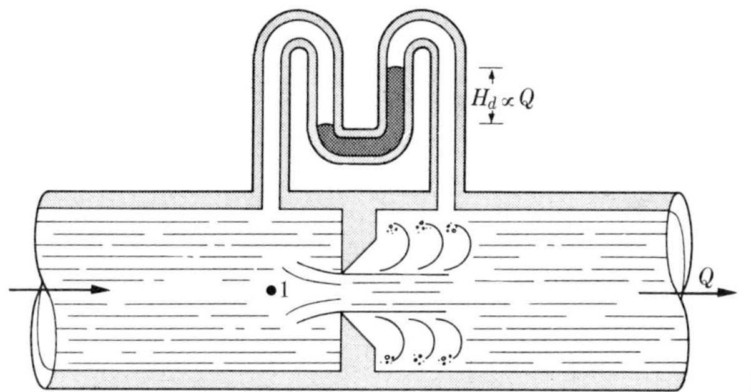

Fig. 8–6

Orifice meters are easier to make and install than venturi meters; however, they do cause a higher pressure loss in the system. Because of the difficulty encountered in the analysis of systems involving orifices and similar flow devices, much of the technology which has grown up around them is based on empirical knowledge. A large quantity of experimental data has been gathered to characterize orifices and establish such factors as discharge coefficients, etc. In spite of this, to ensure its accuracy every flow meter manufactured must be calibrated specifically for the conditions under which it will be used.

Head Loss in an Orifice

In our earlier discussions we established the idea of velocity coefficient:

$$C_v = v_j/v_t, \qquad \text{or} \qquad v_j = C_v v_t = C_v\sqrt{2gh_n},$$

from which

$$h_n = \frac{1}{C_v^2} \cdot \frac{v_j^2}{2g},$$

where h_n is the net head causing flow. The head remaining in the jet, that is, the velocity head, is $v_j^2/2g$. Then,

head loss = (initial head) − (remaining head),

or

$$H_L = \frac{1}{C_v^2}\cdot\frac{v_j^2}{2g} - \frac{v_j^2}{2g} = \left(\frac{1}{C_v^2} - 1\right)\frac{v_j^2}{2g}. \tag{8–6}$$

Equation (8–6) gives the head loss in terms of velocity head remaining in the jet and the velocity coefficient.

We can also express the head loss in terms of the initial head. From Torricelli's equation, we have

$$v_j^2/2g = C_v^2 h_n.$$

Substitution yields

$$H_L = \frac{1}{C_v^2}\cdot C_v^2 h_n - C_v^2 h_n.$$

Then

$$H_L = (1 - C_v^2)h_n. \tag{8–7}$$

Equation (8–7) is probably the more useful of the two expressions.

EXAMPLE

Assume that we have an orifice installation similar to that shown in Fig. 8–5. The diameter of the pipe is 4 in.; that of the orifice is 2 in.; $C_d = 0.60$. A differential manometer is positioned across the orifice and indicates a differential head of 12.43 in. Hg. What is the flow rate of water through the orifice? Calculate the head loss across the orifice if $C_v = 0.98$.

Solution.

Step 1. Define the parameters in consistent units:

$$A_1 = \pi d_1^2/4 = \pi(4^2/4) = 12.57 \text{ in}^2,$$

$$A_o = \pi d_o^2/4 = 3.1416 \text{ in}^2,$$

$$d_o/d_1 = \tfrac{2}{4} = \tfrac{1}{2}, \qquad (d_o/d_1)^4 = (\tfrac{1}{2})^4 = \tfrac{1}{16}$$

$$C_d = 0.6,$$

$$C_d^2 = 0.36,$$

$$(P_1 - P_2)/\gamma = 12.43 \text{ in. Hg} \times 13.6 = 169 \text{ in. } H_2O.$$

Step 2. Substitute in formula (8–5):

$$Q = C_d A_o \sqrt{2g(P_1 - P_2)/\gamma}\,[1 + \tfrac{1}{2}C_d^2(d_o/d_1)^4],$$

$$\begin{aligned} Q &= 0.6 \times 3.1416\sqrt{772 \times 169}\,(1 + (0.36/2) \times \tfrac{1}{16}) \\ &= 0.6 \times 3.1416 \times 27.8 \times 13 \times 1.011 \\ &= 688 \text{ in}^3/\text{sec} \\ &= 688 \times \tfrac{1}{1728} = 0.399 \text{ ft}^3/\text{sec} \\ &= 0.399 \times 7.49 \text{ gal/ft}^3 = 2.98 \text{ gps} \\ &= 2.98 \text{ gps} \times 60 \text{ sec/min} = 179 \text{ gpm.} \end{aligned}$$

Step 3. Using Eq. (8–7), calculate the head loss:

$$C_v = 0.98, \qquad C_v^2 = 0.96,$$

$$H_L = (1 - C_v^2)h_n,$$

$$h_n = v_1^2/2g + (P_1 - P_2)/\gamma,$$

$$v_1 = Q/A_1 = 1150/12.57 = 91.5 \text{ ips},$$

$$v_1^2/2g = (91.5)^2/772 = 10.85 \text{ in.},$$

$$\begin{aligned} H_L &= (1 - 0.96)(10.85 + 169) \\ &= 0.04 \times 179.85 = 7.2 \text{ in. } H_2O. \end{aligned}$$

This represents about $(7.2/169) \times 100 = 4.26\%$ of the indicated manometer head difference.

IMPORTANT TERMS

Orifice is an opening (with a closed perimeter) in an element of a flow system. Although the usual shape is circular, orifices may take any form.

Round-edged orifice is one in which the upstream edge has a radius.

Sharp-edged orifice is one in which the upstream edge is a knife edge.

Vena contracta is that part of the section where the jet issuing from an orifice ceases to contract; or where, in theory, the jet becomes of constant cross-sectional area.

Coefficient of contraction is the dimensionless ratio of the area of the jet at the vena contracta to the area of the orifice opening.

Coefficient of velocity is the dimensionless ratio of the actual velocity in the jet to the theoretical velocity, as calculated by the Torricelli equation.

Discharge coefficient is a dimensionless number obtained from the product of the coefficient of contraction and the velocity coefficient. It is the constant of proportionality in the expression for calculating the flow through an orifice.

Approach velocity relative to an orifice is the mean velocity of the fluid flowing in the conductor upstream from the orifice.

Orifice meter is a device applied to an orifice so that the pressure drop induced as a function of flow is impressed across a manometer which is calibrated to read in flow rate terms.

PROBLEMS

8–1 Water flows through a sharp-edged orifice 2 in. in diameter, as shown in Fig. 8–7. If the head is 100 ft, what is the flow rate? Neglect losses.

8–2 Given a sharp-edged orifice, such as in Fig. 8–7. The head is 50 ft. Calculate the theoretical and actual velocities.

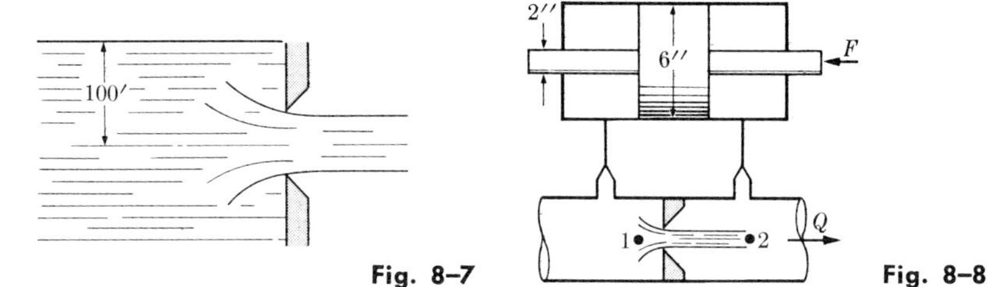

Fig. 8–7

Fig. 8–8

8–3 In a sharp-edged orifice, similar to that in Fig. 8–7, the head is 64 ft. The flow rate, Q, is measured and found to be 846 gpm. Calculate the diameter of the orifice, the jet diameter at the vena contracta, and the location of the vena contracta.

8–4 In actual orifice applications a phenomenon known as "wire-drawing" occurs. Basically, this is the wearing away of the metal at the sharp edge, due to the abrasive action of fluid flowing across the edge. What effect do you think this phenomenon would have on the sharp-edged characteristics of the orifice? How would this affect the orifice coefficients and the accuracy of flow measurements made with such an orifice?

8–5 Discuss the significance of the vena contracta relative to the flow rate through an orifice. What effect does the fluid head have on the vena contracta?

8–6 Consider a flow system similar to that shown in Fig. 8–3. The pressure $P_a = 5$ psi; the height $h = 36$ ft; the jet issues into the atmosphere through an orifice 2 in. in diameter. Calculate the flow rate of water through the orifice. Neglect losses.

8–7 In a flow system similar to that of Problem 8–6, $h = 49$ ft, the orifice diameter is 2 in, and 6.25 gpm of water flows through the orifice into the atmosphere. What is P_a? Calculate the actual and theoretical velocities. What is the diameter at the vena contracta and where is its approximate location?

8–8 Under the conditions of Problem 8–7, what would be the flow rate if the fluid were hydraulic oil? sea water?

8–9 Using the flow system of Problem 8–6, calculate the flow rate of phosphate ester hydraulic fluid into an ambient pressure, $P_B = 2$ psi.

8–10 Consider a flow system such as is shown in Fig. 8–4. The pressure at point 1 is 75 psi; an orifice 3 in. in diameter is located at the end of a 5-in. pipe. Calculate the flow rate of glycerine discharging through the orifice into the atmosphere.

8–11 Consider a flow system similar to that of Problem 8–10; 8 gpm of sea water is flowing through a sharp-edged orifice (diameter 2 in.) in the end of a pipe 4 in. in diameter. Calculate the height to which the fluid will rise in the piezometer tube shown in Fig. 8–4.

8–12 Gasoline flows through a $1\frac{1}{2}$-in. orifice inserted in a 4-in. pipe, as shown in Fig. 8–5. The flow rate is 12 gpm. Calculate the approximate pressure drop across the orifice.

8–13 If a manometer were placed across the orifice of Problem 8–12, what would be the differential head, H_d, when mercury is the gage fluid?

8–14 Returning to Problem 8–1, calculate the head loss across the orifice. Now recalculate the flow rate incorporating this loss in the equation. What percent of error in the flow rate calculation was introduced by neglecting the head loss in Problem 8–1?

8–15 Repeat the procedure of Problem 8–14 for the conditions of Problem 8–2. What is the percent of error with respect to the velocity calculations?

8–16 Calculate the head loss which would actually be encountered in Problem 8–10. Recalculate the flow rate and determine the percent of error introduced by neglecting the head loss.

8–17 Using the conditions of Problem 8–13, make a number of calculations of H_d for several flow rates and plot a curve of H_d vs. Q. If the orifice of Problem 8–13 were to be used as a flow meter, what would be the significance of this curve? How would the neglect of head loss across the orifice affect the accuracy of this characteristic curve?

8–18 Consider the orifice meter of Problem 8–17. Discuss the effect on accuracy of using the meter for a fluid other than that for which it was calibrated.

8–19 Figure 8–8 shows a technique used in industry for control purposes. An orifice is placed in a flow line. A pressure drop is developed across the orifice by the flowing fluid. This drop is impressed across a piston in a cylinder to provide a control force, F, proportional to the flow rate, Q^2. In the illustration, the pipe is 2 in. in diameter, the orifice is 1 in. in diameter, and the fluid is water-glycol hydraulic fluid. Calculate $P_1 - P_2$ for several values of Q between 1 and 20 gpm, and plot a characteristic curve of Q vs. F.

CHAPTER 9

Nozzles, tubes, and similar flow devices

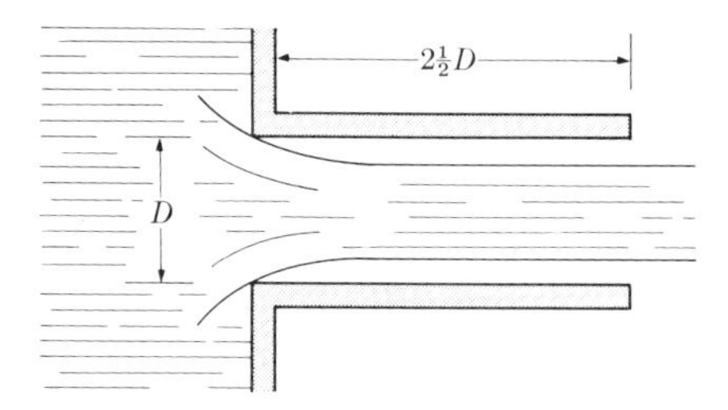

Fig. 9–1

Our discussion of orifices and orifice flow characteristics in Chapter 8 can be extended to other types of flow devices.

Standard Short Tube

The standard short tube is illustrated in Fig. 9–1. It is a circular conduit $2\frac{1}{2}$ times its diameter in length and has a square cornered entrance edge.

Consider what happens when flow is started suddenly. The fluid may spring clear of the wall of the short tube, in which case the tube will act like a sharp-edged orifice. This is shown in Fig. 9–1. On the other hand, if flow starts more gradually, the jet will undergo a contraction similar to that observed in an orifice. A short distance downstream the jet will reexpand to the full diameter of the tube. Thus the coefficient of contraction for a short tube running full is $C_c = 1.00$. It has been determined that the velocity coefficient for such a short tube is $C_v = 0.82$. Since the discharge coefficient is the product of the two, we have $C_d = 1 \times 0.82 = 0.82$. Thus the flow discharged by a standard short tube is about one-third greater than that discharged by a sharp-edged orifice of the same diameter

($C_d = 0.60$). Figure 9–2 illustrates a short tube running full. It has been demonstrated that the pressure head at the vena contracta is

$$P_1/\gamma = -0.82h_n.$$

If the fluid is water, P_1/γ cannot be less than -34 ft. Thus h_n cannot be greater than 41.5 ft if the tube is to run full.

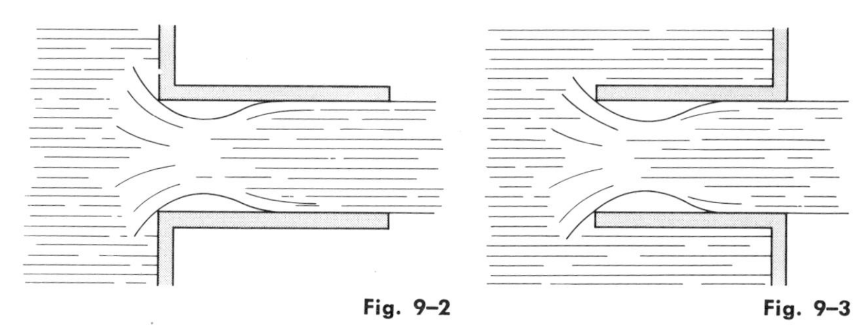

Fig. 9–2 Fig. 9–3

Head Loss

The head loss can be calculated using the same form as Eq. (8–6):

$$H_L = \left(\frac{1}{C_v^2} - 1\right)\frac{v^2}{2g}.$$

When $C_v = 0.82$ is substituted, we have

$$H_L = 0.50\,\frac{v^2}{2g}. \tag{9–1}$$

One of the most significant points is that the entrance to a pipe set flush with the wall of a reservoir acts like a standard short tube with respect to the fluid entering the pipe.

Reentrant Tubes

Rather than being flush with the surface, as is the case with a standard short tube, a reentrant tube projects into the reservoir, as shown in Fig. 9–3. Flow characteristics are similar to those encountered in the short tube, except for the fact that the jet contracts more. When the reentrant tube is running full, $C_c = 1$. For a reentrant tube, $C_v = 0.75$. Thus $C_d = 0.75$. The head loss for a reentrant tube can be calculated as above:

$$H_L = 0.8(v^2/2g). \tag{9–2}$$

Borda's Mouthpiece

A Borda mouthpiece is a special case of a reentrant tube in which a very thin-walled tube projects back into a reservoir a length equal to the diameter of the tube (Fig. 9–4). It can be demonstrated analytically that under ideal conditions

$$C_v = 1 \quad \text{and} \quad C_c = 0.50.$$

Under practical conditions when $C_v = 0.98$, then $C_c = 0.52$.

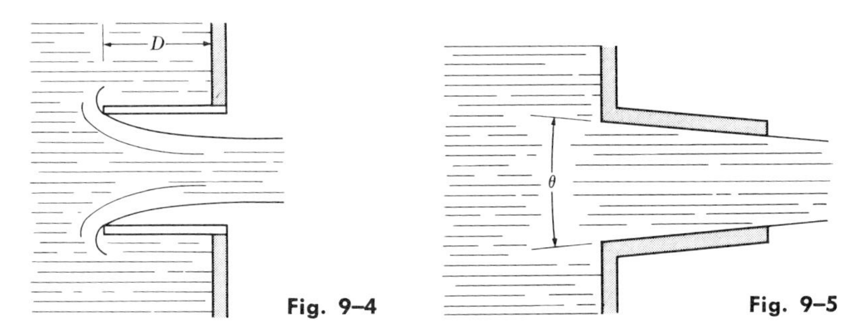

Fig. 9–4 Fig. 9–5

Converging and Diverging Tubes

A *converging* tube is one whose cross section is not constant throughout its length but is gradually reduced from a larger opening at the reservoir wall to a smaller opening at the free end (Fig. 9–5).

The flow characteristics of a converging tube are a function of the included angle, θ. As θ increases, C_c decreases; the limiting value is 0.61 when $\theta = 180°$, that is, for an orifice. Similarily, C_v decreases as θ decreases. Because the value of θ can range from 0° to 180°, the reader should consult a suitable handbook to find the corresponding coefficient values. A diverging tube is the opposite of a converging tube. As shown in Fig. 9–6, the cross sectional area of a diverging tube increases as the tube proceeds

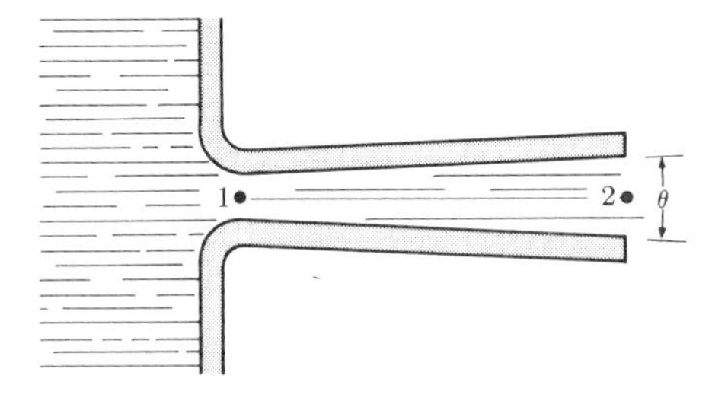

Fig. 6–1

outward from the wall of a reservoir. Since a sharp edge is not practical for a diverging tube, the entrance is rounded.

The flow characteristics of *diverging* tubes are based on empirical data. It has been shown that the greatest flow rate can be attained with an included angle $\theta = 5°$, and a length equal to nine times the throat diameter.

Nozzles

To increase the velocity of the fluid a nozzle, which is a "converging tube," is attached to the end of a pipe. The velocity increases in accordance with the principle of flow continuity for an incompressible fluid: $Q = A_1v_1 = A_2v_2 = \cdots$. See Fig. 9–7. During this process some of the potential energy due to pressure is converted to kinetic energy. Thus the increase in velocity is accompanied by a decrease in pressure.

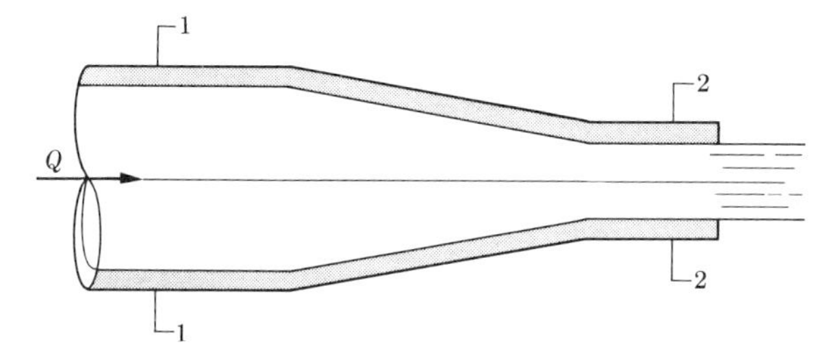

Fig. 9–7

The head loss in a nozzle can be calculated by means of Eq. (8–6), where the specific value of C_v will have been empirically determined for the nozzle configuration under consideration. Since the normal range of values lies between 0.98 and 0.96, this loss is usually neglected.

Diffusers

A *diffuser* is the converse of a nozzle. It is essentially a "diverging tube," whose purpose is to increase the pressure in a fluid stream by reducing the flow velocity (Fig. 9–8). When the fluid stream is slowed down, some of the kinetic energy in the stream is converted to potential (pressure) energy.

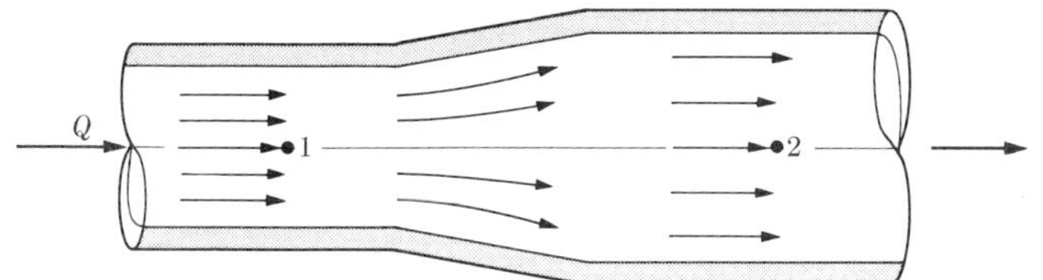

Fig. 9–8

Fig. 9–9

Hydraulic components are used extensively in the precision operation of this adjustable center driller for structural steel shapes. The machine has 40 top drivers and 40 adjustable center spindle heads. It is numerically controlled, again demonstrating the flexibility of fluid power.

The pressure recovery in a practical diffuser will not quite reach that calculated by means of Bernoulli's equation alone, since there are some losses in the nonideal case. The efficiency of a diffuser can be approximated by the formula

$$e_d = \frac{P_2 - P_1}{\frac{1}{2}\rho v_1^2[1 - (A_1/A_2)^2]}, \tag{9–3}$$

where P_1 = pressure at section 1,
P_2 = pressure at section 2,
ρ = density of fluid, slugs,
v_1 = velocity at section 1,
A_1 = cross-sectional area at section 1,
A_2 = cross-sectional area at section 2.

Flow and Control Devices

Many of the characteristics of fluid power and process flow control or measurement mechanisms, for applications like those of Fig. 9–9, can be determined by approximation to orifices, tubes, nozzles, or combinations of these basic devices. As is so often the case in engineering practice, quantitative data can only be obtained empirically. Thus much of the following information is a summary of data obtained by testing the various devices. Table 9–1 summarizes the basic equations of flow and pressure drop for simple orifices and nozzles. The curves in Table 9–2 represent values of C_d for the various types of orifices and nozzles.

TABLE 9–1*

Flow	*Pressure drop*	*Flow*	*Pressure drop (measured across flow device)*
$\frac{\text{ft}^3}{\text{sec}}$	ft	$Q = d_o^2 C_d \sqrt{H_L}\, 4.38 \times 10^{-2}$	$H_L = \frac{Q^2}{d_o^4 C_d^2} 5.21 \times 10^2$
$\frac{\text{lbs}}{\text{sec}}$	psi	$W = d_o^2 C_d \sqrt{\gamma\, \Delta P}\, 5.25 \times 10^{-1}$	$P = \frac{W^2}{d_o^4 C_d^2 \gamma} 3.63$
gpm	psi	$q = d_o^2 C_d \sqrt{\Delta P/\gamma}\, 2.36 \times 10^2$	$P = \frac{q^2 \gamma}{d_o^4 C_d^2} 1.8 \times 10^{-5}$

* Courtesy of *Product Engineering* and Crane Company.

TABLE 9–2*

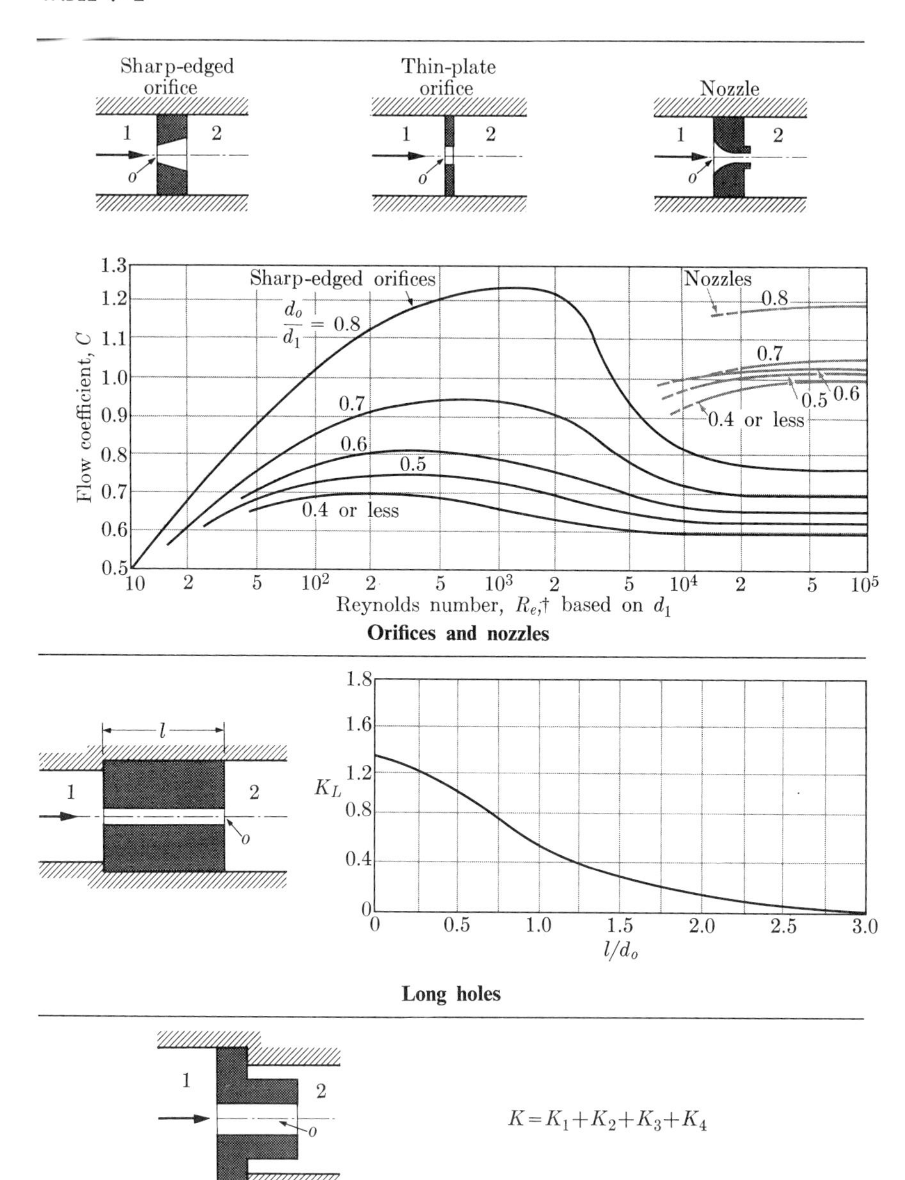

Pipe diameter change (long hole)

* Courtesy of *Product Engineering* and Louis Dodge, Consulting Engineer.
† Refer to Reynolds Number, p. 136.

TABLE 9–2 (cont.)

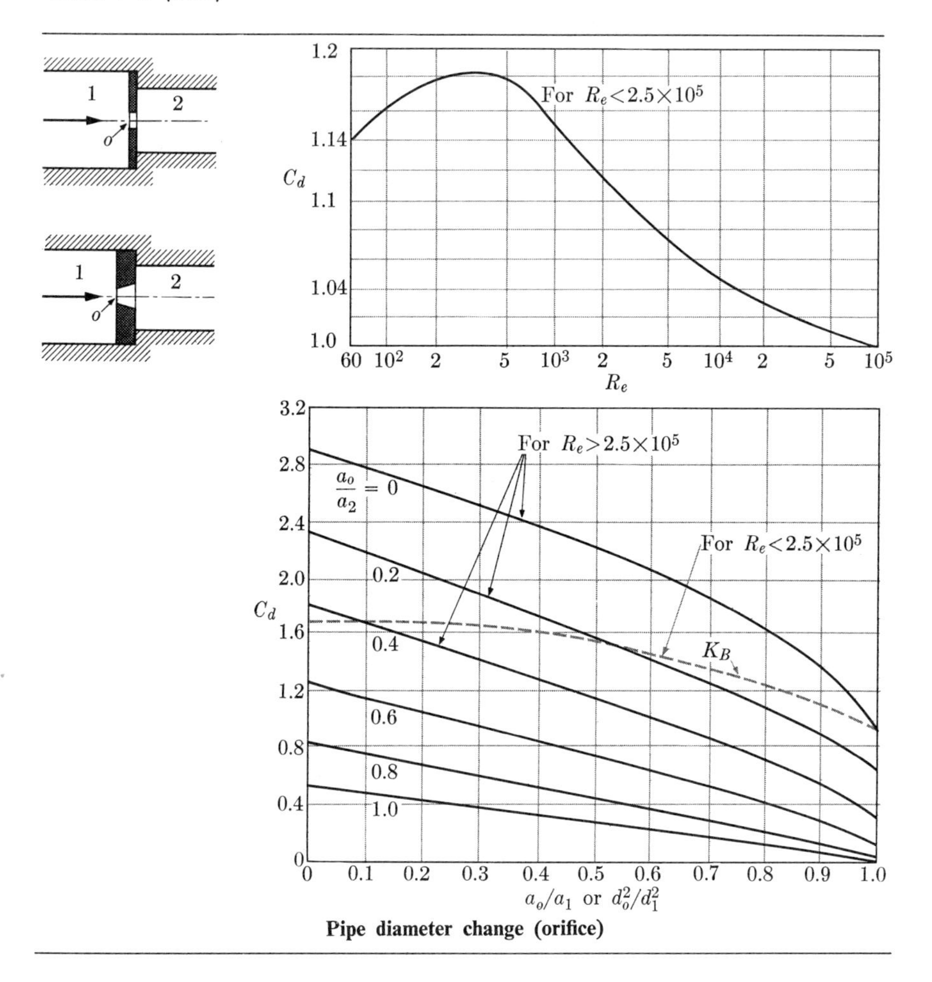

Pipe diameter change (orifice)

The head loss, H_L, across a flow device, such as an orifice, nozzle valve, etc., can be expressed in the form $H_L = Kv^2/2g$, that is, as a function of the velocity head multiplied by some characteristic constant K. These K-values are determined empirically for the particular flow device in question. The velocity v is the average velocity just before and just after the restriction. For orifices and holes, *except* sharp-edged orifices and nozzles, the velocity is that which is determined from the principles of flow continuity, $Q = Av$. For valves and slots, v is the approach velocity at the inlet port area. Some cases involve a so-called friction factor considered in Chapter 14 on pipe flow. Table 9–3 lists characteristic K-values for a number of flow devices frequently encountered in industry.

TABLE 9–3*

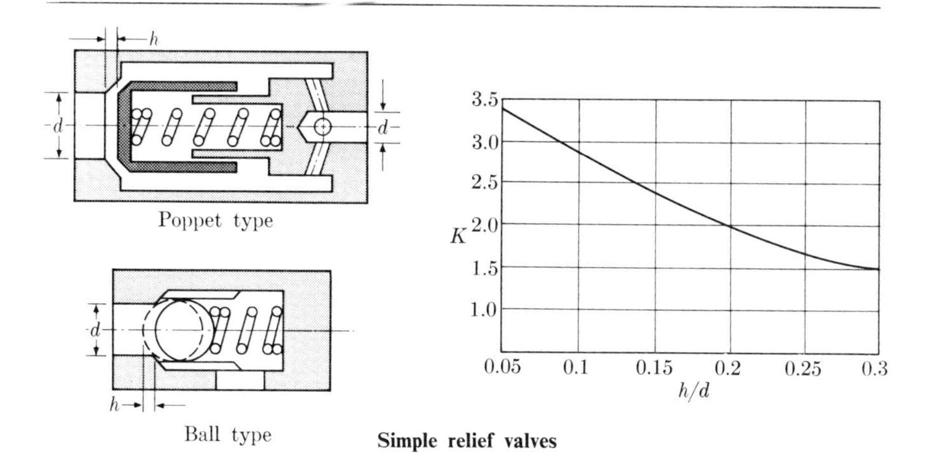

Simple relief valves

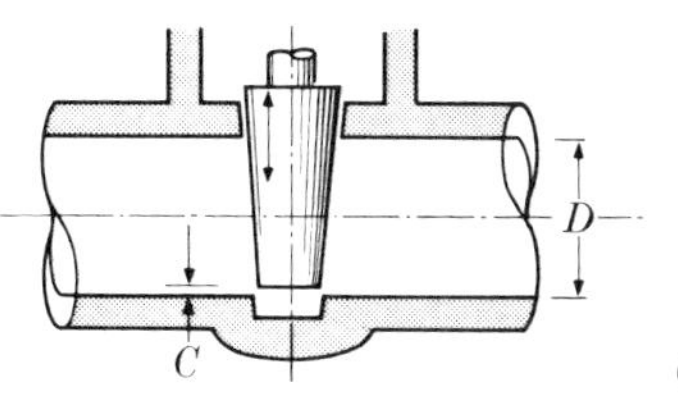

C/D	1	0.9	0.8	0.7	0.6	0.5	0.4	0.3	0.2	0.1
K	1.3	1.6	2.0	3.0	4.5	6.2	10	20	50	200

Gate valve

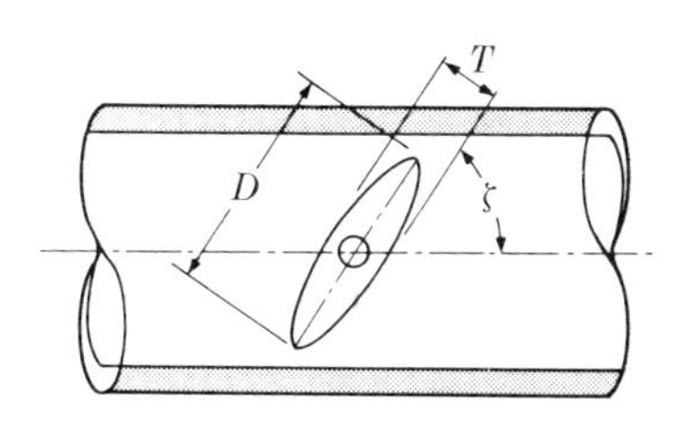

T/D	0.5	0.4	0.3	0.2	0.1
K	2	1	0.5	0.2	0.1

Various wafer thicknesses, assuming fully open ($\zeta = 0$) in a closed system. For free discharge, add 1.0 to K-values.

ζ	0	10	20	30	40	50	60	70	80
K	1	1.2	1.8	4	12	35	145	600	closed

All for $T/D = 0.4$

Butterfly valve

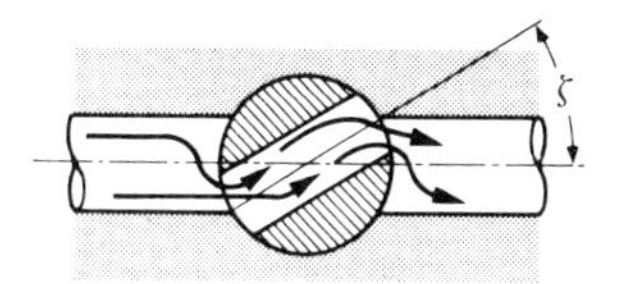

ζ	0	10	20	30	40	50	60	70	75
K	0	0.3	1.6	5.5	18	54	210	1000	closed

Ball valve

* Courtesy of *Product Engineering* and Louis Dodge, Consulting Engineer.

TABLE 9–3 (cont.)

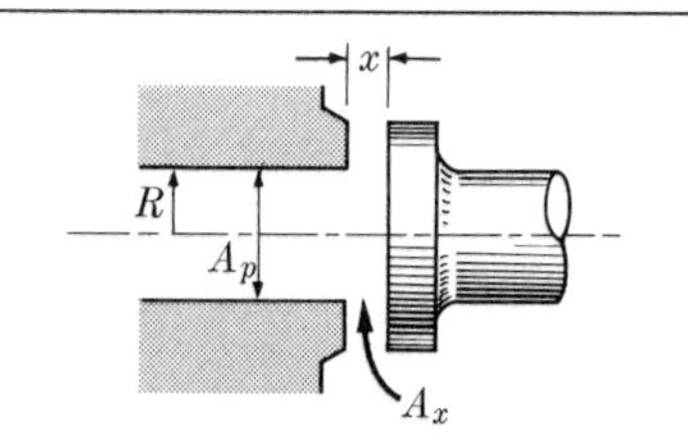

Orifice area:

$A_x = 2\pi R x$

$A_x = A_p$ when $x = R/2$

Wetted perimeter: $L_w = 4R$

Resistance coefficient:

$K = 1.3 + 0.2(A_p/A_x)^2$

Disc valve

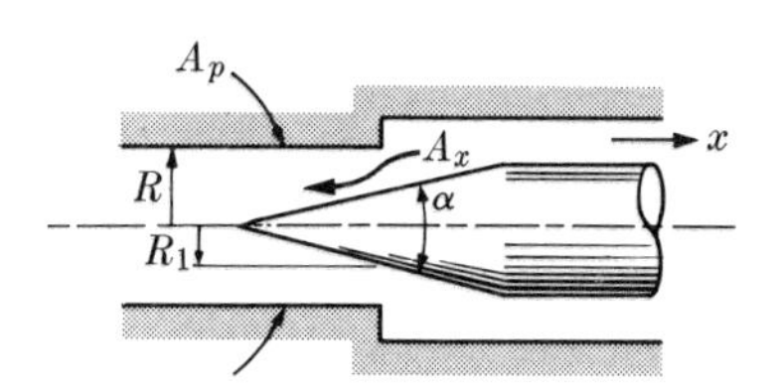

$$A_x = \pi\left(2Rx \tan\frac{\alpha}{2} - x^2 \tan^2\frac{\alpha}{2}\right)$$

$A_x = 0$, when $x = 0$

$$L_w = 2\pi\left(2R - x \tan\frac{\alpha}{2}\right)$$

$$K = 0.5 + 0.15\left(\frac{A_p}{A_x}\right)^2, \text{ when}$$

$0.10A_p < A_x < A_p$

Needle valve

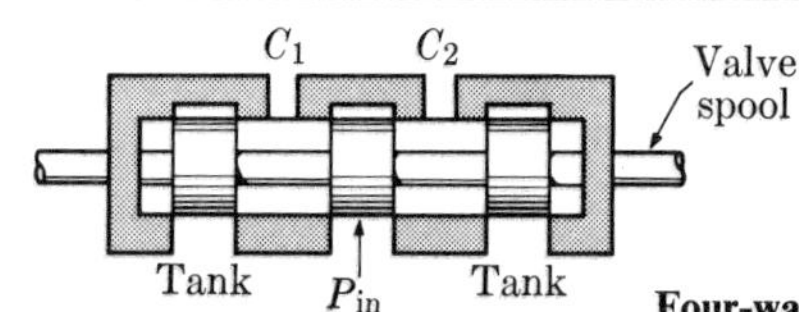

$K = 3$ to 5.5 when valve is fully open

Four-way valve

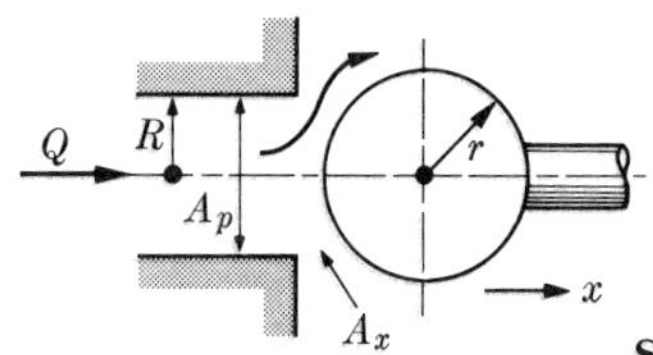

$A_x \approx 1.5R\pi x$

A_x is valid when $r \approx 1.3R$

$L_w \approx 4\pi R$

$K \approx 0.5 + 0.15(A_p/A_x)^2$

Simple ball valve

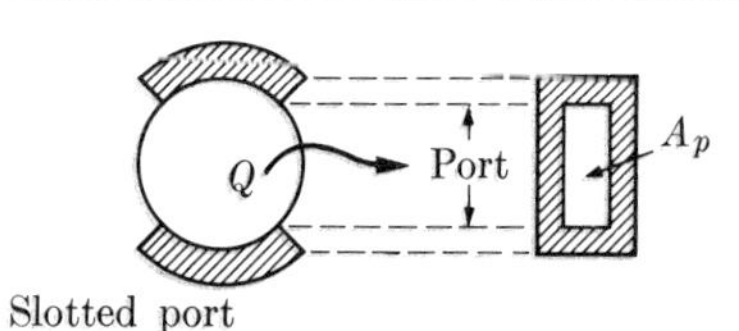

Slotted port

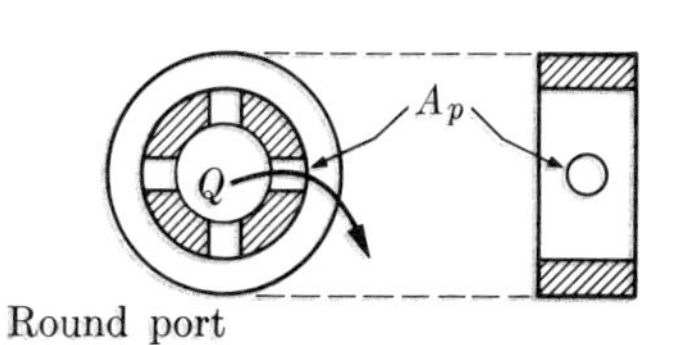

Round port

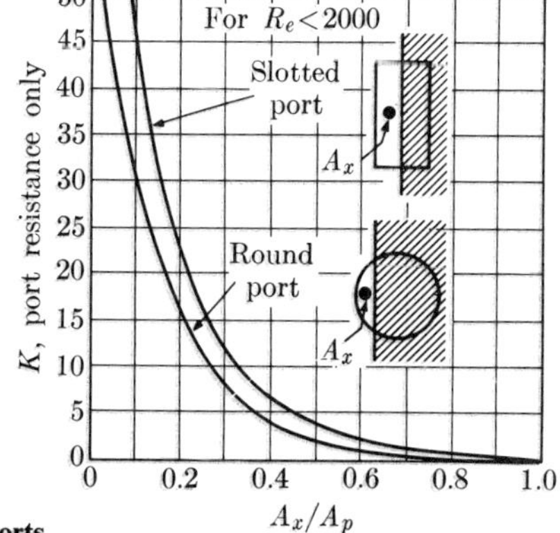

Typical ports

TABLE 9–3 (cont.)

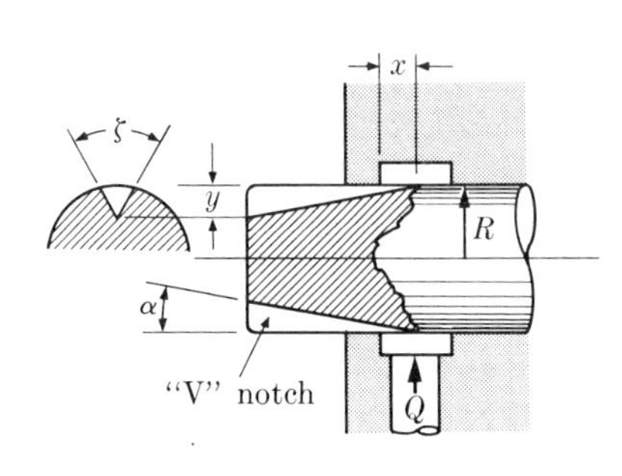

$A_x \approx \frac{\zeta}{2} y^2$, ζ in radians

$y = x \tan \alpha$ when $\zeta = \frac{\pi}{3}$

$A_x = N \frac{\pi}{6} x^2 \tan^2 \alpha$ N = number of notches

Wetted perimeter:

$L_w = \pi y \frac{\zeta}{180} + 2y$

$K \approx \frac{400}{N_R}$ when $N_R < 150$ (Reynolds' number)

$K \approx \frac{10}{N_R} 0.25$ when $150 < N_R < 2000$

Notches in plunger

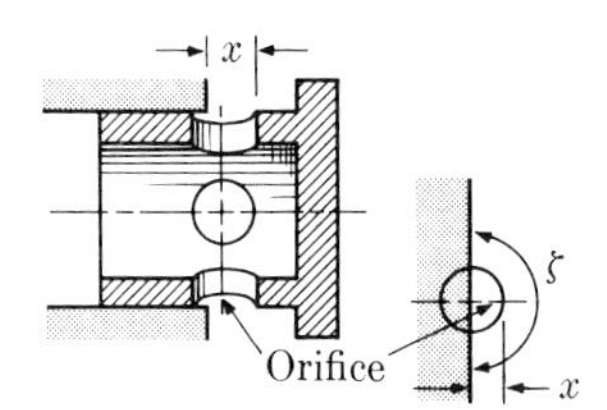

$A_x = NR^2 \cos^{-1} \frac{R-x}{R} - (R-x)\sqrt{2Rx - x^2}$

$A_x = \frac{1}{2} R^2 (\zeta - \sin \zeta)$, ζ in radians

$L_w = R\left(\zeta + 2 \sin \frac{\zeta}{2}\right)$

K: same as curves under "Typical ports" above

Partial orifice

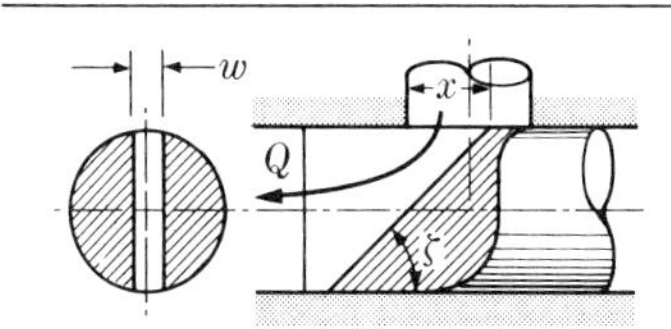

K: same as slotted port curve under "Typical ports" above

Ramp slot

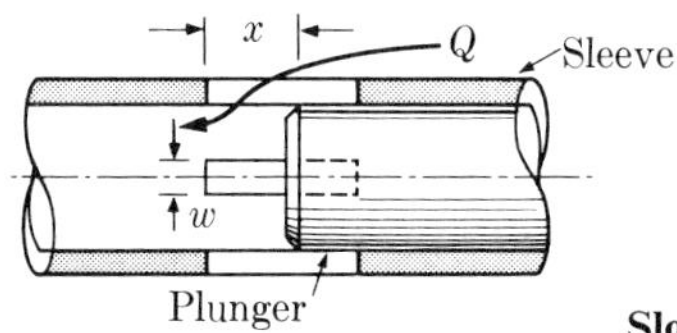

$A_x = Nwx$ N = number of slots

$L_w = 2N(w + x)$

K: same as slotted port curve under "Typical ports" above

Slotted sleeve

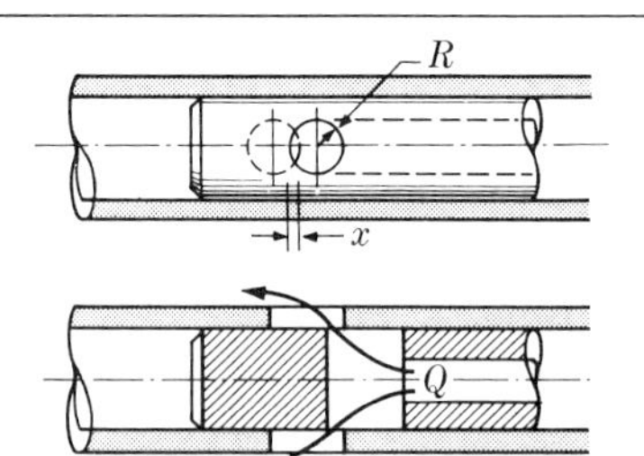

$$A_x = 2\left[R^2 \cos^{-1}\left(\frac{R - x/2}{R}\right) - \left(R - \frac{x}{2}\right)\sqrt{2R\frac{x}{2} - \left(\frac{x}{2}\right)^2}\right]$$

K: same as round port curve under "Typical ports" above

Intersecting holes

TABLE 9–3 (cont.)

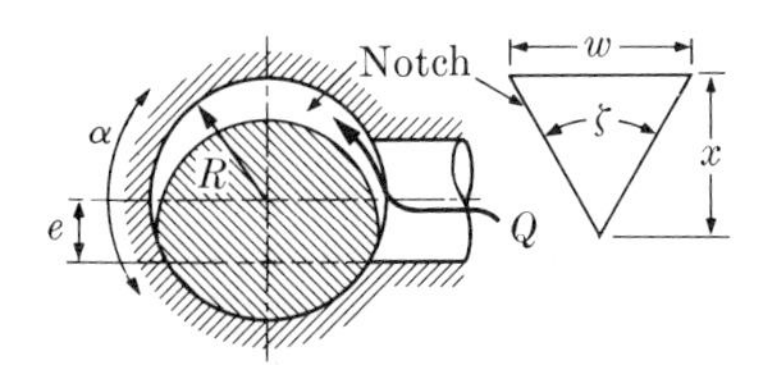

Notch section height:

$$x = \sqrt{e^2 + R^2 - 2eR\cos\alpha} - R$$

$$x = 0 \quad \text{when } \cos\alpha = \frac{e}{2R}$$

Area: $A_x = w\dfrac{x}{2}$ $\quad w = 2\left(\tan\dfrac{\zeta}{2}\right)x$

K: same as "Notches in plungers" above

Rotary notch

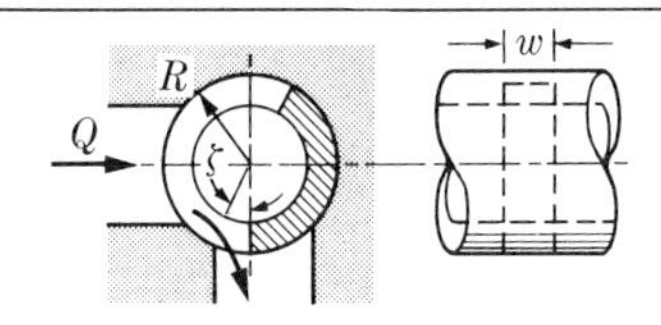

$A_x = Rw\zeta$

$L_w = 2(R\zeta + w)$

K: same as slotted port curve under "Typical ports" above

Rotary slot

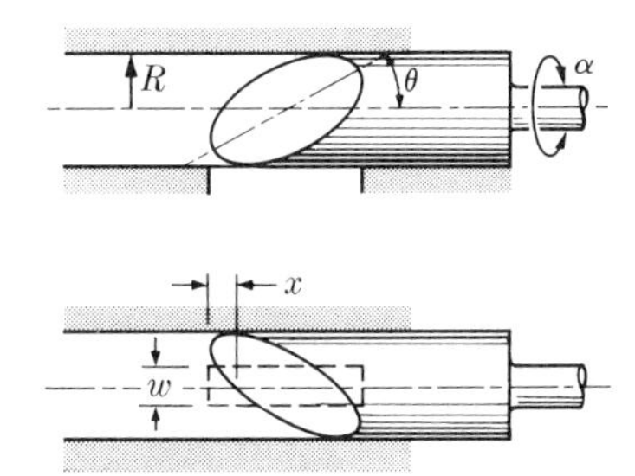

$A_x = w(1 - \cos\alpha)R\cot\theta$

$L_w = 2(x + w)$

$x = R\cot\theta(1 - \cos\alpha)$

K: same as slotted port curve under "Typical ports" above

Rotary wedge

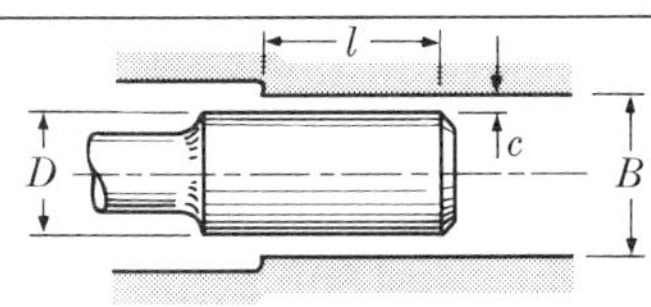

$$\Delta P = \frac{12Q_{\text{in}^3}\mu_m l}{C_{\text{rad}}^3\, d_m\pi}$$

Annular clearance

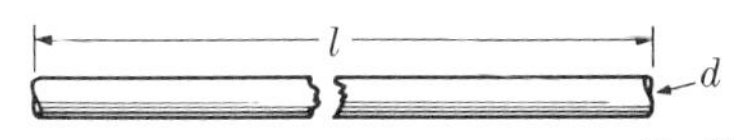

$$\Delta P = \frac{Q_{\text{in}^3}\mu l}{d^4} \times 40.75$$

Capillary tubing

A helical conductance path machined into the periphery of a cylindrical member, then pressed into a cylindrical bore.

$$\Delta P = \frac{Q_{\text{in}^3}\mu l}{a^2}\cdot 32$$

where $a = w \cdot y$
w = width of groove
y = depth

Screw path

A sintered or otherwise porous material placed in a fluid conductor to produce a pressure drop.

$$\Delta P = CQ_{\text{in}^3}\mu \cdot \frac{l}{a_2}$$

Porous plug

Flow Coefficient

Some manufacturers of fluid-power control valves use a coefficient, C_v, to describe flow characteristics of valves. The relationship $Q_{gpm} = C_v\sqrt{p}$ describes how the flow coefficient is used. The relationship between K-factor and C_v is $K = 1460a^2/C_v^2$, where a is the flow area, in^2.

The preceding discussion has covered factors describing flow characteristics of the devices most frequently used in industrial controllers and fluid-power mechanisms. Also illustrated is the application of the basic orifice equation to special cases.

IMPORTANT TERMS

Standard short tube is a tube in which the length is $2\frac{1}{2}$ times the diameter.

Reentrant tube is a tube which, rather than projecting outward from the reservoir, projects back into it.

Borda's mouthpiece is a special case of reentrant tube, in which a very thin tube projects back into the reservoir over a length equal to the diameter of the tube.

Converging tube is a tubular conductor in which the cross-sectional area diminishes from the entrance to the exit section of the tube.

Diverging tube is a tube in which the cross-sectional area increases from the entrance to the exit section.

Nozzle is a flow device which is a converging tube attached to the end of a pipe.

Diffuser is essentially a diverging tube intended to increase the pressure of a fluid by reducing the stream velocity.

K-factor is used in characterizing flow through various devices. It is considered to be the constant of proportionality between the head loss (pressure drop) across the flow device and the velocity head of the stream.

PROBLEMS

9–1 Water discharges through a standard short tube 3 in. in diameter, as illustrated in Fig. 9–1. If the tube is running full, what is the flow rate under 100 ft of head? Neglect losses.

9–2 Repeat Problem 9–1, but include the head loss. What percent of error was introduced by neglecting the loss originally?

9–3 Ethyl alcohol discharges through a reentrant tube similar to that shown in Fig. 9–3. The tube is $1\frac{1}{2}$ in. in diameter and under 49 ft of head. Calculate the flow rate for fully developed flow (a) neglecting losses, (b) including losses. What is the obvious implication?

9–4 Water issues from a nozzle such as that shown in Fig. 9–7. The diameter at section 1–1 is 4 in.; at 2–2 it is $1\frac{1}{2}$ in. If the pressure at 1–1 is 150 psi, what is the velocity of the free jet? If $Q = 35$ gpm; what is the velocity? Assume $\theta = 5°$.

9–5 Consider a diffuser as illustrated in Fig. 9–8. Let $A_1 = 0.7854$ in^2, and $A_2 = 7.07$ in^2. Water flows at a rate of 15 gpm. Pressure at point 1 is 100 psi. Using Bernoulli's equation, calculate the pressure at point 2.

9–6 Calculate the efficiency of the diffuser in Problem 9–5, if the pressure at p_2 is 100.2 psi. What percent of error would be introduced by using Bernoulli's equation alone?

9–7 A water-in-oil emulsion fluid flows at 10 gpm through a single poppet-type relief valve with an inlet port diameter $d = \frac{3}{4}$ in. The poppet lifts a distance $h = 0.1125$ in. off its seat. Calculate the pressure drop across the valve.

9–8 A gate valve has an inlet diameter of 2 in. It is open a distance $C = 0.8$ in. If MIL-5606 hydraulic fluid flows through the valve at a rate of 80 gpm, what is the pressure drop across the valve?

9–9 How much pumping horsepower would be needed to overcome the loss across the valve of Problem 9–8?

9–10 A disk valve has an inlet port diameter of $1\frac{1}{2}$ in. When 75 gpm of hydraulic oil pass through the valve, it opens $\frac{3}{16}$ in. Calculate the pressure drop across the valve and the added pumping horsepower required to overcome it.

9–11 The relationship of horsepower consumed to heat generated is 1 hp = 42.44 Btu/min. How much heat would be generated under the conditions of Problem 9–9? How much heat would be generated under the conditions of Problem 9–10?

9–12 The specific heat of a substance is defined as the number of Btu required to raise the temperature of 1 lb of the substance 1°F. The specific heat of MIL 5606 is about 0.465 Btu/lb/°F; that of a typical mineral-base hydraulic oil is about the same. Using the calculations of Problem 9–11 and assuming that all the heat generated is used to raise the temperature of the oil, determine the temperature rise of the hydraulic fluids passing through the valves.

9–13 A typical application of a four-way (direction-control) valve is shown in Fig. 9–10. Hydraulic fluid flows in through the port P_{in} to the cylinder port C_1 and then to the cylinder. The return flow comes from the rod end of the cylinder, passes through C_2, and out through the tank port. Assume that the port P_{in} is $\frac{3}{4}$ in. in diameter and that 30 gpm of hydraulic oil is flowing in. Assume an average K-factor. Calculate the pressure drop across the valve from P_{in} to C_1, and from C_2 to the tank. What would be the anticipated temperature rise of the fluid?

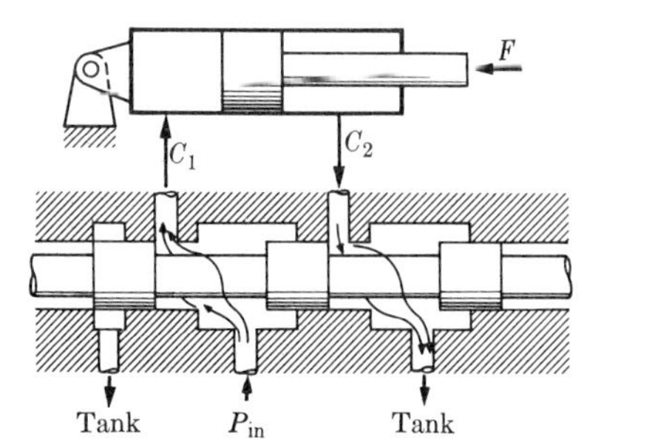

Fig. 9–10

9–14 In Chapter 8, we developed an equation relating the characteristics of an orifice, flow rate, and pressure (or head) drop across the orifice by

$$Q = C_d A_o \sqrt{2gh_n} = C_d A_o \sqrt{2g(\Delta P/\gamma)}. \tag{1}$$

For a given orifice this could be written in the form

$$Q = K_o \sqrt{\Delta P}. \tag{2}$$

In this chapter we discussed the calculation of head loss (pressure drop) across various flow devices, using an expression of the form

$$H_L = K \frac{v^2}{2g} \equiv \frac{\Delta P}{\gamma}.$$

Note the similarity of the two expressions. What function of ΔP is the flow rate Q through an orifice? Using considerations of flow continuity, convert Eq. (2) into one, in Q and ΔP, of the form of Eq. (1) (for orifice flow). In this new equation, what function of ΔP is Q? What is the significance of these relationships of ΔP and Q?

9–15 From your solution to Problem 9–14, it should be apparent that many of the orifice-like flow devices discussed in Chapter 9 are used to control flow rate as a function of ΔP. This is accomplished by varying the "orifice" opening to control ΔP and thereby Q. The needle valve is a typical example. Consider one with an inlet port diameter $(2R)$ of $\frac{1}{2}$ in. The included angle of the cone, $\alpha = 10°$, and $x = \frac{1}{4}$ in. If the pressure drop across the needle valve is 250 psi, what is the flow rate Q of hydraulic oil through the valve?

9–16 The rotary notch is widely used in fluid-power flow-control valves as an orifice of variable area. The port diameter is $\frac{1}{2}$ in.; $R = \frac{1}{2}$ in.; $e = \frac{1}{8}$ in.; $\alpha = 30°$; $\zeta = 45°$. If the pressure drop across the orifice is 500 psi, what is the flow rate Q?

CHAPTER 10

Flow under conditions of changing head

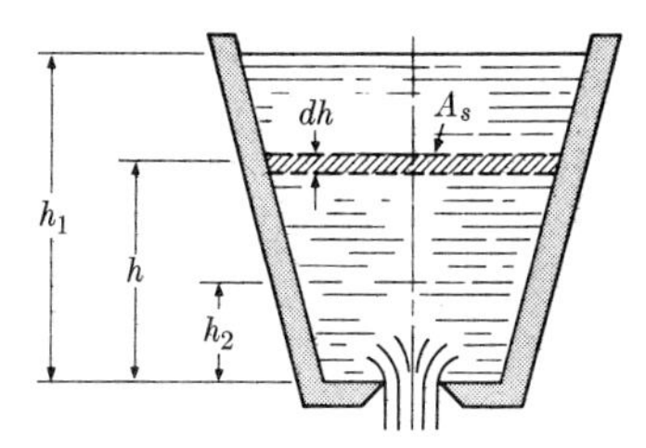

Fig. 10–1

Concept of Falling Head

In our previous discussions of flow we have assumed that a *constant head* provides the energy required to cause fluid flow. Thus, when we derived the expression for Torricelli's equation in Chapter 7, we used a constant head, h.

It is obvious that many flow systems do exhibit this condition. However, in practice, we frequently begin with a full reservoir and let the fluid run out through an orifice or similar flow device. Thus the level of fluid in the reservoir falls and the head goes with it. We must find a method for coping with this situation.

A typical case is illustrated in Fig. 10–1. We have a reservoir containing an incompressible fluid and an orifice in the bottom of the reservoir through which the fluid is flowing. At some time t_1, we measure the level of the surface of the fluid above the orifice, that is, the head on the orifice, and determine that it is h_1.

There are several phenomena which we might wish to evaluate. (1) We might want to find the flow rate at time t_1; or (2) we might want to know the rate at which the surface of the fluid is falling in the reservoir;

or (3) we might want to calculate the time it would take for the liquid level to fall from the initial height of h_1 to some second level, h_2. As a matter of fact, the latter condition is the one which usually is of interest to us.

The method of attacking the problem consists of writing equations for flow rate and volume discharged during a given period of time, and solving them for that time.

Step 1. At any intermediate head h between h_1 and h_2, the equation for flow through the orifice would be

$$Q = C_d A_o \sqrt{2gh}.$$

Step 2. In the time increment dt, the volume discharged at the above flow rate would be

$$dV = Q\,dt = C_d A_o \sqrt{2gh}\,dt. \tag{10–1}$$

Step 3. In the same time increment dt, the level (or head) will have dropped by an increment dh.

Step 4. We can write a second equation for the discharged volume, using the drop in head, dh, and the area of the surface of the liquid at that time, A_s. (Note that A_s must not be confused with A_o, the area of the orifice.) Thus

$$dV = A_s\,dh. \tag{10–2}$$

Step 5. Equating these two expressions for the volumetric increment, we have

$$dV = A_s\,dh = C_d A_o \sqrt{2gh}\,dt.$$

Step 6. Solving for time, we obtain

$$dt = A_s\,dh / C_d A_o \sqrt{2gh}.$$

Step 7. In order to evaluate this expression, the surface area must be expressed as a function of the head h, $A_s = f(h)$, and this expression must be integrated between the limits of h_1 and h_2:

$$t = \int_{h_2}^{h_1} \frac{f(h)\,dh}{C_d A_o \sqrt{2gh}}. \tag{10–3}$$

Equation (10–3) gives the basic form of the solution to the problem of flow under a falling head. As you can see, the problem now becomes one of expressing the surface area in terms of head and evaluating the integral.

EXAMPLE

Consider a cylindrical tank, one of the most common types of reservoirs encountered in engineering work. (See Fig. 10–2.) Fluid is discharged through an orifice in the bottom of the reservoir. Determine an expression for the time required to lower the head from h_1 to h_2.

Solution.

Step 1. The flow rate through the orifice is

$$Q = C_d A_o \sqrt{2gh}.$$

Step 2. The incremental volume is

$$dV = A_s\, dh = C_d A_o \sqrt{2gh}\, dt.$$

Step 3. Equating and solving for t, we get

$$t = \int_{h_2}^{h_1} \frac{A_s\, dh}{C_d A_o \sqrt{2gh}} = \frac{A_s}{C_d A_o \sqrt{2g}} \int_{h_2}^{h_1} h^{-1/2}\, dh.$$

In this particular case, A_s is constant.

Step 4. Thus we have

$$t = \frac{2A_s}{C_d A_o \sqrt{2g}} (\sqrt{h_1} - \sqrt{h_2}). \qquad (10\text{–}4)$$

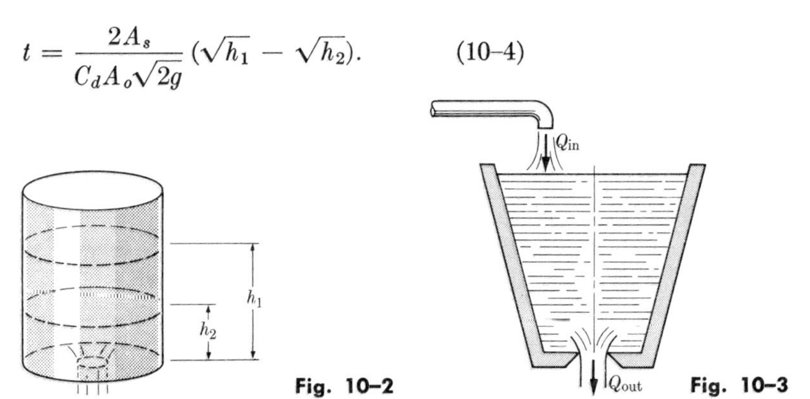

Fig. 10–2 Fig. 10–3

Flow In and Out

In some applications encountered in industry, there may be an inflow of liquid at the same time that outflow occurs through the orifice, as shown in Fig. 10–3.

There are two possible cases.

Case 1. The inflow is greater than the outflow:

$Q_{\text{in}} > Q_{\text{out}}.$

Case 2. The outflow is greater than the inflow:

$Q_{\text{out}} > Q_{\text{in}}.$

The approach to the solution of the problem is the same as that described previously; that is, we write two equations for the discharged volume, equate them, and solve for the time.

Case 1.

$$dV = A_s\, dh;$$

1) here the incremental volume is expressed in terms of area and head.

$$dV = (Q_{\text{in}} - Q_{\text{out}})\, dt;$$

2) in this case, dV is expressed in terms of flow rate and time. Then

$$t = \int_{h_1}^{h_2} \frac{A_s}{Q_{\text{in}} - Q_{\text{out}}}\, dh$$

$$= \int_{h_1}^{h_2} \frac{A_s}{Q_{\text{in}} - C_d A_o \sqrt{2gh}}\, dh.$$

It can be seen that Q_{in} must be known to evaluate the expression. Note that in this case, the head would be rising since more liquid is flowing in per unit time than is flowing out.

Case 2.

$$dV = A_s\, dh;$$

1) here we have a minus sign, since the head is falling in this case.

$$-dV = (Q_{\text{out}} - Q_{\text{in}})\, dt;$$

2) in this case, $Q_{\text{out}} > Q_{\text{in}}$ so that Q_{in} is subtracted from it to yield a positive volume of fluid. Note that the head would be falling in this case.

The equations are solved in the same manner as in Case 1.

EXAMPLE

Consider a cylindrical tank similar to that of Fig. 10–2. Under what *constant head*, h_c, would the orifice discharge the same volume of liquid as would be discharged under a head falling from 9 ft to 4 ft?

Solution.

Step 1. Volume discharged under falling head = volume discharged under constant head.

Step 2.

$$t = -\frac{A_s}{C_d A_o \sqrt{2g}} \int_{h_1}^{h_2} \frac{1}{\sqrt{h}}\, dh.$$

Step 3.

$$t = \frac{2A_s(\sqrt{h_1} - \sqrt{h_2})}{C_d A_o \sqrt{2g}}.$$

Step 4. Multiply and divide by $(\sqrt{h_1} + \sqrt{h_2})$.

Step 5. Then

$$t = \frac{A_s(h_1 - h_2)}{\frac{1}{2}C_d A_o(\sqrt{h_1} + \sqrt{h_2})\sqrt{2g}}.$$

Step 6. Note that

$$\tfrac{1}{2}C_d A_o(\sqrt{h_1} + \sqrt{h_2})\sqrt{2g} = \tfrac{1}{2}(Q_1 + Q_2),$$

or the average flow rate under varying flow conditions.

Step 7. Then

$$\tfrac{1}{2}C_d A_o \sqrt{2g}\,(\sqrt{h_1} + \sqrt{h_2})t = C_d A_o \sqrt{2gh_c}\, t.$$

Step 8. Furthermore,

$$\tfrac{1}{2}(\sqrt{9} + \sqrt{4}) = \sqrt{h_c}.$$

Step 9. Hence

$$h_c = 6.25 \text{ ft}.$$

IMPORTANT TERMS

Constant head is that condition in a reservoir where the height of liquid above the discharging orifice remains constant.

Falling head is that condition in a reservoir in which the height of liquid above the orifice is dropping because of the outflow of liquid through an orifice.

PROBLEMS

10–1 Consider a cylindrical tank similar to that shown in Fig. 10–2. If the head on a 3-in. orifice is held constant at 49 ft, what is the flow rate through the orifice?

10–2 A cylindrical tank is 10 ft in diameter. With no inflow, calculate the time required to lower the head of water in the tank from 36 ft to 9 ft through a 3-in. orifice. Determine the average flow rate, Q_a.

10–3 In the system of Problem 10–2, what inflow rate would be required to maintain the head constant at 36 ft?

10–4 The initial head on an orifice was 16 ft and the final head was 4 ft. Calculate the constant head h under which the same orifice would discharge the same volume of fluid in an equal time interval.

10–5 Given a tank which is shaped like the frustum of a cone (Fig. 10–1). Let h_1 be 15 ft, the diameter at the top surface be 10 ft, and that of the bottom be 5 ft. Determine the size of the sharp-edged orifice in the bottom which will cause the tank to empty in 10 min. [*Hint:* Express the surface area as a function of the head h. Also note that the total time equals $\int dt$.]

10–6 A rectangular tank 5 ft wide is divided as shown in Fig. 10–4. A sharp-edged orifice 6 in. in diameter is located as indicated. How long will it take for the water surfaces to level out? [*Hint:* The difference in level at any given time on the orifice is h. A drop in level on the left side is related to a rise on the right side by the ratio of the areas of the two surfaces.]

10–7 A cylindrical tank 10 ft in diameter is connected near the bottom to a second cylindrical tank 5 ft in diameter by a pipe which acts essentially as a short tube. If the pipe (short tube) is 4 in. in diameter and the 5-ft tank is filled with oil ($S_g = 0.8$) to a height of 20 ft above the connecting pipe, how long will it take for the oil level to equalize in the two tanks? At what height above the pipe will it equalize?

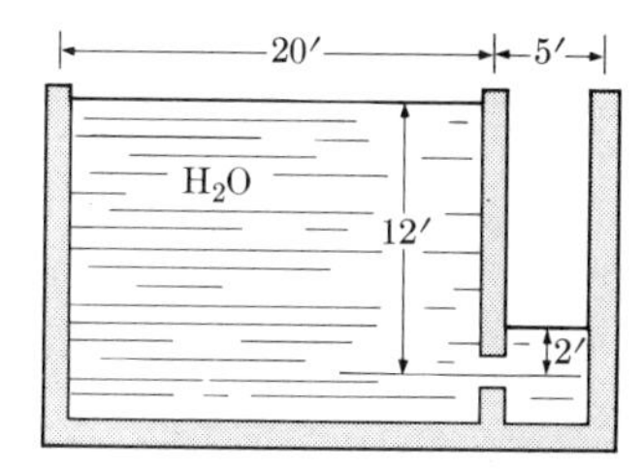

Fig. 10–4

10–8 The S.S. Milwaukee is an oil tanker with bunkers (tanks) having trapezoidal cross sections. The bunkers are 100 ft across at the top, 60 ft across at the bottom, 40 ft high and 40 ft long. Each tank is connected to its neighbor by a 12-in. pipe which is equipped with a gate valve to isolate the tanks. Assume that bunker 1 is filled to $\frac{3}{4}$ of its depth with oil ($S_g = 0.85$). Bunker 2 is empty. Now the 12-in. gate valve is opened halfway. How long will it take for the oil level to equalize in the two tanks?

10–9 Set up the equation to determine inflow rate required to double the time calculated in Problem 10–2 to lower the head from 36 ft to 9 ft.

10–10 Given a cylindrical standpipe 5 ft in diameter, with a 2-in. orifice in the bottom, set up the equation to determine inflow rate required to *raise* the head of water from 5 ft to 10 ft in 20 sec.

CHAPTER 11

Introduction to hydrodynamics

So far we have developed some concepts of fluid mechanics which are somewhat isolated from the environments in which they are actually applied. We can legitimately do this during the development stage, but at some time we must attempt to relate these flow systems to their surroundings. In this and the following chapters we will move from the "ideal" situation referred to in Chapter 1 to a broader and more generalized application of the principles discussed earlier.

Impulse and Momentum

A fluid stream, particularily a liquid, is a system involving a moving mass. From our earlier discussions, we recall the simple expression for flow rate: $Q = A \cdot v$. If A and v are given in units of linear measure, Q evolves as volumetric flow rate. If we multiply Q by the density ρ, we get the mass flow rate.

Since we have a moving mass which is undergoing changes in velocity, Newton's second law of motion describes the situation:

$$F = ma.$$

This expression indicates that when a constant force is applied to a mass it will be uniformly accelerated; F is pounds force when m is given in slugs and a is given in ft/sec^2. Conversely, when a mass is uniformly accelerated it exerts a force against that body which causes the motion.

If we let the acceleration a equal the final velocity minus the initial velocity divided by the time interval over which the velocity change took place, the equation takes the form

$$F = m\frac{(v_2 - v_1)}{t}, \tag{11-1}$$

where

v_1 = initial velocity, v_2 = final velocity, and t = time.

The above equation can be written in the form

$$Ft = m(v_2 - v_1),$$

where the product Ft is called the *impulse*, and the product mv is the *momentum*. The relationship states that the *impulse* equals the *change in momentum*. This concept will be used later in the development of the expression for the reaction of a jet of fluid on a surface, such as a turbine blade.

Equation (11–1) could also be written in the following form:

$$F = \frac{m}{t}(v_2 - v_1).$$

We can see that the quantity m/t is the mass flow rate. If we define a quantity M_1 such that $M_1 = m/t$, Eq. (11–1) can be rewritten as

$$F = M_1(v_2 - v_1),$$

which represents the reaction force a surface would have on the fluid stream by virtue of a change in velocity, $v_2 - v_1$. Conversely, the fluid jet would exert a force on the surface of equal magnitude, but opposite in sense, that is, $-F$. Thus the equation for the force of a jet impinging on a surface is

$$F = M_1(v_1 - v_2). \tag{11-2}$$

Velocity

Velocity is a vector quantity, that is, one which has both magnitude and direction. Thus the change in velocity indicated in Eq. (11–2) can be a change in magnitude, as shown in Fig. 11–1(a), or it can be a change in

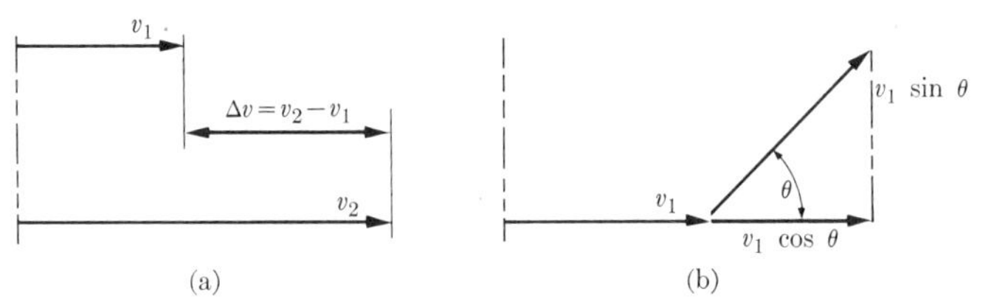

Fig. 11–1

direction with magnitude remaining constant, as in Fig. 11–1(b), or it could be both. Note that in the second case, the change in direction actually encompasses a velocity change in two directions. In the x-direction Δv is $v_1 - v_1 \cos \theta$. In the y-direction Δv is $v_1 \sin \theta$, since there was no initial velocity component in the y-direction.

We can measure velocity in two modes:

1) relative to some absolute reference, such as the earth's surface (at least for us on the earth), in which case we have what is called *absolute velocity*.
2) with reference to some other moving object, in which case we speak of *relative velocity*.

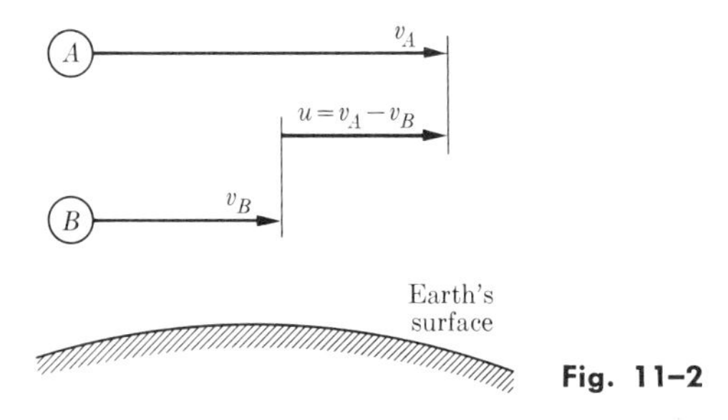

Fig. 11–2

This is illustrated schematically in Fig. 11–2. Objects A and B both have absolute velocity with respect to the earth. They also have relative velocity with respect to each other, such that

$$u = v_A - v_B,$$

where u = relative velocity.

Fluid Jet Impingement, Moving Object

We are going to consider an important concept involving the reaction (force) of a jet of fluid impinging on a surface. A "pedagogical fantasy" will be used to make the explanation easier to follow. Assume that we have a single vane moving in a straight line through space with a velocity v_v, as

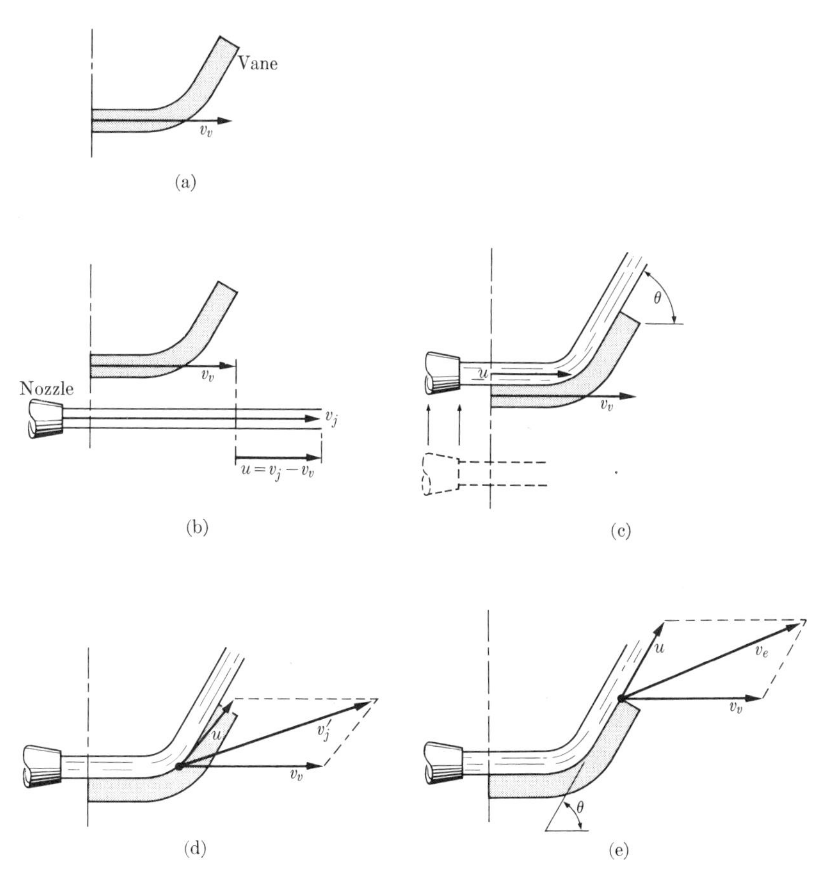

Fig. 11–3

shown in Fig. 11–3(a). This is an absolute velocity. Now assume that we also have a nozzle from which a jet of liquid is issuing into space with a jet velocity of v_j(absolute). The path of the vane and the jet are parallel, as shown in Fig. 11–3(b). We can see that the relative velocity between the vane and the jet is $u = v_j - v_v$. At some instant in time, t_0, we move the nozzle up so that the jet impinges on the vane, as shown in Fig. 11–3(c). The velocity of the jet of fluid moving over the surface of the vane is u, the relative velocity. If we assume a frictionless flow of fluid over the vane, this velocity will remain constant. At some intermediate point on the vane, the velocity relationship will look like that shown in Fig. 11–3(d). At each such point the jet will have a new absolute velocity, v_j', which will be the vector sum of u and v_v. Figure 11–3(e) shows the relationship as the jet leaves the vane. The jet will be tangent to the surface and will therefore have an exit angle θ.

The final velocity in the x-direction is $u \cos \theta$. Final velocity in the y-direction is $u \sin \theta$. Thus the change in velocity in the x-direction is the initial velocity u minus the final velocity, or

$$\Delta v_x = u - u \cos \theta = u(1 - \cos \theta).$$

Similarily, the change in velocity in the y-direction is

$$\Delta v_y = 0 - u \sin \theta = -u \sin \theta.$$

Apparent Flow Rate

It would appear as though all we had to do would be to substitute these values into Eq. (11–2) and calculate the reaction force. However, such a procedure would give an erroneous answer. We must first consider the actual M_1, the mass flow rate per unit time. If we go back to Fig. 11–3(b), we can ascertain that the mass flow rate issuing from the nozzle is $Q_m = \rho A_j v_j$. But is this the flow rate per unit time which the vane "sees"? No, it is not! Recall that the velocity of the jet *relative* to the vane is u, not v_j. Therefore the flow rate which the vane experiences is

$$Q' = A_j u \rho.$$

Or, if we write density in terms of the specific weight and g, we have

$$Q' = A_j u \gamma / g = M_1.$$

Now we can proceed with Eq. (11–2):

$$F_x = M_1 \, \Delta v_x = \frac{A_j u \gamma}{g} u(1 - \cos \theta) = \frac{A_j u^2 \gamma}{g} (1 - \cos \theta)$$

$$= \frac{A_j \gamma (v_j - v_v)^2}{g} (1 - \cos \theta). \tag{11–3}$$

Also

$$F_y = -M_1 v_y = -\frac{A_j \gamma (v_j - v_v)^2}{g} \sin \theta. \tag{11–4}$$

It is obvious that the vane of our illustration could not remain in straight-line motion in space with these two forces acting on it. However, if we picture the vane as one on the periphery of a turbine wheel, the explanation becomes more plausible. In such a case, the x-reaction would be that which causes the turbine to revolve, while the y-reaction would show up as a thrust load on the wheel, which would have to be absorbed by the turbine bearings.

Work Done

We are interested in the work done on the vane by the impinging jet. Recall that work done can be calculated by taking the product of the force and the distance through which it moves:

$$G = F \times d.$$

Thus

$$G_x = \frac{A_j\gamma(v_j - v_v)^2}{g}(1 - \cos\theta)v_v \quad \text{per unit time.}$$

It can be seen that the distance per unit time is

$$d = v_v t = v_v \cdot 1 = v_v.$$

We can determine the conditions under which the maximum amount of work can be done by differentiating the above expression and setting it equal to zero. Then finding the roots, we have

$$dG/dv_v = \frac{\gamma A_j(1 - \cos\theta)}{g}(v_j^2 - 4v_jv_v + 3v_v^2) = 0,$$

from which

$$v_v = v_j \qquad \text{and} \qquad v_v = v_j/3.$$

Since no work is done at all when $v_v = v_j$, the maximum work would be done on the vane by the jet when $v_v = v_j/3$.

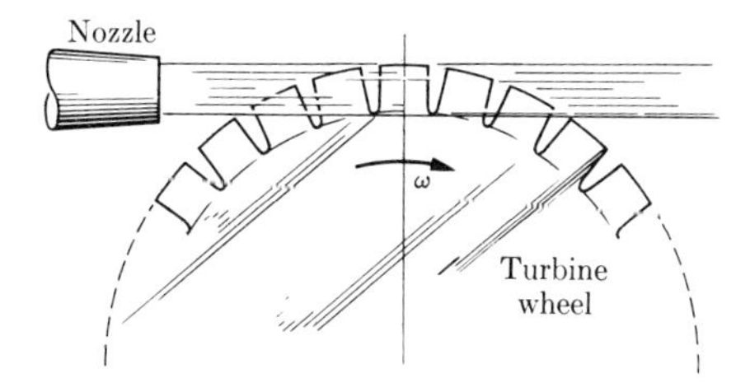

Fig. 11–4

Since the concept of a single vane in space is an imaginary device used to simplify the discussion of jet reaction, the above calculation of work has no real significance other than for purposes of demonstrating the method. Let us consider a fluid turbine containing a series of vanes arranged around the periphery of the wheel in such a way that the entire jet is always impinging on one or more vanes (Fig. 11–4). In this case the actual jet flow rate would also be the relative flow rate, and $M_1 = \gamma A_j v_j/g$.

Then

$$F_x = \frac{\gamma A_j v_j}{g}(v_j - v_v)(1 - \cos\theta),$$

and

$$G = \frac{\gamma A_j v_j}{g}(v_j - v_v)(1 - \cos\theta).$$

By a method similar to that above, we can show that the maximum work is done when $v_v = v_j/2$.

EXAMPLE

A vane moves through space with a velocity of 30 fps. A nozzle 2 in. in diameter delivers a jet with a velocity of 50 fps such that it impinges on the vane. The exit angle of the vane is 60°. The fluid is oil with a specific gravity of 0.8. Calculate the reaction forces on the vane due to the jet.

Solution.

Step 1. From the previous derivation we know that the reaction force in the x-direction will be

$$F_x = \frac{A_j \gamma (v_j - v_v)^2}{g}(1 - \cos\theta).$$

Step 2. Evaluate parameters:

a) $A_j = \pi d^2/(144 \times 4) = \pi/144 = 0.0218 \text{ ft}^2$,

b) $\gamma = 62.4 \text{ lb/ft}^3 \times 0.8 = 50 \text{ lb/ft}^3$,

c) $g = 32.2 \text{ ft/sec/sec}$,

d) $(v_j - v_v) = (50 - 30) = 20 \text{ fps} = u$,

e) $1 - \cos\theta = 1 - \cos 60° = 1 - 0.5 = 0.5$.

Step 3. Substitute and calculate:

$$F_x = \frac{50 \text{ lb/ft}^3 \times 0.0218 \text{ ft}^2 \times 20^2 \text{ft}^2/\text{sec}^2}{32.2 \text{ ft/sec}^2} \times 0.5$$

$$= 6.76 \text{ lb}.$$

Step 4. Check dimensionally:

$$\text{lb} = \frac{\text{lb} \times \text{ft}^4 \times \text{sec}^2}{\text{ft}^4 \times \text{sec}^2} = \text{lb}.$$

Step 5. Similarly:

$$F_y = \frac{-A_j\gamma(v_j - v_v)^2}{g}\sin\theta$$

$$= \frac{-0.0218 \times 50 \times 400 \times 0.866}{32.2} = -11.7 \text{ lb.}$$

Jet Impingement, Stationary Object

Note that when the jet impinges on a stationary vane, then $v_v = 0$, and the relative velocity is

$$u = v_j - v_v = v_j.$$

Figure 11–5 shows a jet impinging on a stationary flat plate at an angle θ with the horizontal. If surface friction between the plate and the jet of fluid is neglected, the total change in momentum of the jet is perpendicular to the plate. This is true in spite of the fact that the jet splits into two parts, one moving up along the plate and the other moving down. Thus:

$$F = M_1 v_j \sin\theta. \tag{11–5}$$

It can be seen that $v_j \sin\theta$ is the velocity change and that M_1 is, as before, the mass flow rate.

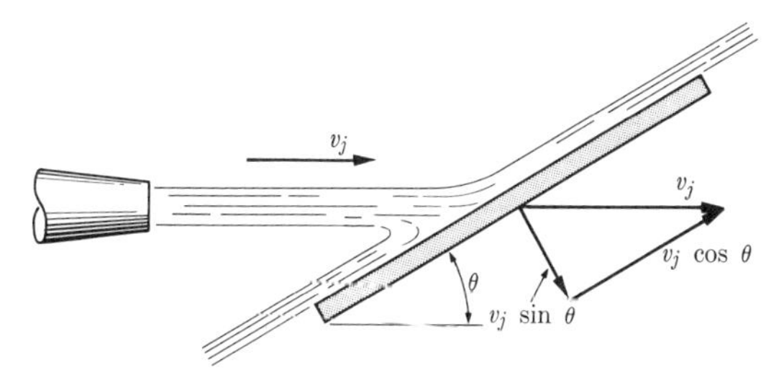

Δv(perpendicular to plate) $= v_j \sin\theta - 0 = v_j \sin\theta$ **Fig. 11–5**

Equation (11–5) is based on the previous derivations predicated on conservation of momentum. That is, under ideal flow conditions, the momentum of the flowing stream prior to the change in direction is equal to the momentum of the stream after the change. If we denote by Q_u the part of the stream flowing up along the surface of the plate and by Q_d the part flowing downward along the plate, then it follows from the principle of flow continuity that the flow of the jet impinging on the plate, Q_j, equals the sum of the split flows, or

$$Q_j = Q_u + Q_d.$$

Because there is no force in the direction parallel to the surface of the plate, there is no change in momentum in that direction. Thus the fluid momentum in the original jet equals the fluid momentum along the plate, or

$$M_1 v_j \cos\theta = Q_u \rho v_j - Q_d \rho v_j,$$

or, since $M_1 = Q_j \rho$,

$$Q_j \rho v_j \cos\theta = Q_u \rho v_j - Q_d \rho v_j,$$

from which

$$Q_u = (Q_j/2)(1 + \cos\theta), \qquad Q_d = (Q_j/2)(1 - \cos\theta).$$

EXAMPLE

A jet of oil ($S_g = 0.8$) flowing at a rate of 5 gpm from a nozzle 1 in. in diameter impinges on a plate inclined at 60° to the horizontal (see Fig. 11–5). What is the force reacting on the plate and what are the flows along the surface of the plate?

Solution.

Step 1. Reduce the parameters to common units:

a) $5 \text{ gpm} \times 231 \text{ in}^3/\text{gal} = 1155 \text{ in}^3/\text{min} = 19.25 \text{ in}^3/\text{sec}$,

b) $v = Q/A_j = 19.25/0.7854 \text{ in}^2 = 24.5 \text{ ips}$,

c) $\sin 60° = 0.866, \quad \cos 60° = 0.500$,

d) $$M_1 = \rho Q_j = \frac{0.8 \times 0.0361 \text{ lb/in}^3}{386 \text{ in/sec}^2} \times 19.25 \text{ in}^3/\text{sec}$$
$$= 0.00144 = 14.4 \times 10^{-4},$$

where $\gamma = 0.8 \times 0.0361 \text{ lb/in}^3$,

$\rho = \gamma/g$,

$g = 32.2 \text{ ft/sec}^2$,

$= 386 \text{ in/sec}^2$.

Step 2.

$$F = M_1 v_j \sin\theta = 14.4 \times 10^{-4} \times 24.5 \times 0.866 = 3.05 \times 10^{-2} \text{ lb}.$$

Step 3.

$$Q_u = (Q_j/2)(1 + \cos\theta) = (19.25/2)(1 + 0.500) = 14.43 \text{ in}^3/\text{sec.}$$

Step 4.

$$Q_d = (Q_j/2)(1 - \cos\theta) = (19.25/2)(1 - 0.500) = 4.82 \text{ in}^3/\text{sec.}$$

IMPORTANT TERMS

Impulse is the product of a force and the time through which it acts, $F \cdot t$.

Momentum is the product of the mass and its velocity, $m \cdot v$.

Velocity is a vector quantity which describes the rate of change of position of a body, and gives both its magnitude (speed) and direction.

Relative velocity is the velocity of one body measured with respect to a second body.

Absolute velocity is the velocity of a body measured with respect to a fixed reference, such as the earth's surface.

Apparent flow rate is the rate at which fluid flows relative to some moving body rather than to an absolute reference.

PROBLEMS

11–1 A jet of water issues from a nozzle 2 in. in diameter, as shown in Fig. 11–5. The flow rate is 200 gpm. The jet impinges on a stationary flat plate inclined at an angle of 45° to the centerline of the jet. What force is exerted on the plate by the jet?

11–2 A jet of water from a nozzle 1 in. in diameter flows at a rate of 60 gpm. It impinges on a flat plate perpendicular to the centerline of the jet. What is the reaction force of the jet on the plate?

11–3 A curved vane similar to that in Fig. 11–3 is held stationary. The exit angle of the vane is 60°. A jet of sea water issues from a nozzle 2 in. in diameter at a rate of 150 gpm. Determine the reaction forces on the structure holding the vane, due to impingement of the jet.

11–4 A curved vane (Fig. 11–3) moves in a straight line with an absolute velocity of 15 fps. The exit angle is 45°. A jet of alcohol issues from a nozzle (diameter 1 in.) at a rate of 90 gpm in a direction parallel to the motion of the vane so as to impinge on it. Calculate the reaction of the fluid jet on the vane.

11–5 A jet of hydraulic oil issues from a nozzle (diameter 1 in.). A pitot tube is placed in the free jet, and the oil rises to a height of 1.55 ft in the vertical leg of the tube. The jet impinges on a curved vane moving at an absolute velocity of 5 fps. The exit angle is 60°. Calculate the reaction of the fluid jet on the vane.

11–6 A water jet issues from a nozzle (1 in. in diameter) at a rate of 48.9 gpm. A pitot tube is placed in the free jet, and the water rises to a height of $74\frac{1}{2}$ in. in the tube. The jet impinges on a curved vane with an exit angle of 45°. A force transducer measures the reaction force, which is found to be 5.96 lb in the direction of the jet. Determine the absolute velocity of the vane.

11–7 A curved vane moves linearly with an absolute velocity of 12 ft/sec. A jet of sea water issues from a nozzle and impinges on the vane. A pitot tube is placed in the free jet, and the sea water rises to a height of 47.8 in. in the vertical leg. The exit angle of the vane is 60°. The measured reaction on the vane is 0.545 lb in the direction of motion. Determine the exit diameter of the nozzle. What is the flow rate of the sea water?

11–8 A jet of water strikes the fixed surface shown in Fig. 11–6 in such a way that it flows at a rate of 2 ft³/sec along each surface. The initial jet velocity is 60 ft/sec. Neglecting surface friction, determine the x- and y-components of the reaction force.

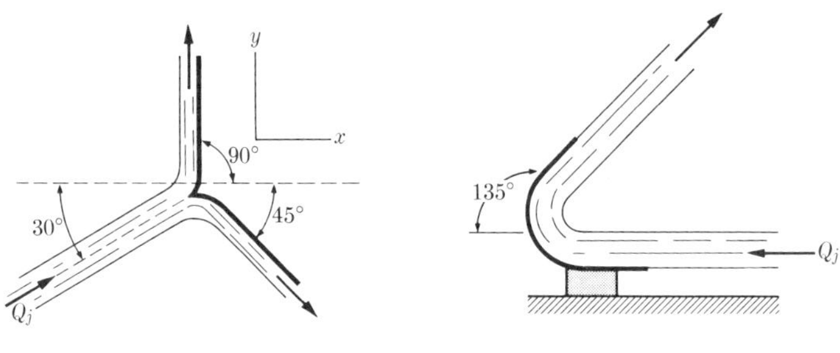

Fig. 11–6 **Fig. 11–7**

11–9 Assume that the surface of Problem 11–8 is moving in the x-direction with a velocity of 20 ft/sec. Determine the reaction-force components under this new condition.

11–10 A jet of sea water impinges on a fixed curved vane which deflects the jet through a 135° angle, as shown in Fig. 11–7. The jet exits from a nozzle 2 in. in diameter at a rate of 300 gpm. Determine the reaction-force components in the x- and y-directions.

11–11 In the example of Problem 11–10, assume that the vane is moving in the x-direction at a velocity of 15.6 ft/sec. Calculate the x- and y-components of the reaction force under these new conditions.

11–12 Consider the system of Problem 11–5, but replace the 1-in. nozzle with a sharp-edged orifice 2 in. in diameter. Using average orifice coefficients where applicable, calculate the reaction force on the vane, the theoretical jet velocity, and the net head required on the orifice to produce the reaction force and the velocity.

11–13 Replace the 1-in. nozzle of Problem 11–6 with a sharp-edged orifice 1 in. in diameter. All other conditions remain the same. Determine the vane velocity. What is the theoretical jet velocity? What head is required on the orifice to produce this flow condition? How does the change from a nozzle to an orifice affect the flow system?

11–14 Calculate the work done by the alcohol jet on the vane of Problem 11–4.

11–15 Determine the work done by the hydraulic oil jet on the vane of Problem 11–5.

11–16 Calculate the work done by the water jet on the surface of Problem 11–9.

11–17 How much work is done by the fluid jet on the curved vane of Problem 11–10?

11–18 Assume the conditions of Problem 11–4, except that the jet impinges on a series of such vanes located on the periphery of a turbine wheel 24 in. in diameter. Calculate the theoretical torque on the wheel and the work done by the jet.

11–19 Repeat Problem 11–18, using the conditions of Problem 11–5.

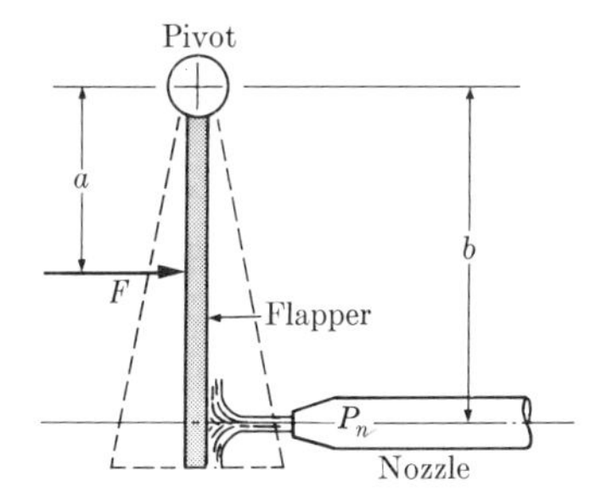

Fig. 11–8

11–20 The *flapper nozzle* shown in Fig. 11–8 is a control device used in fluid-power systems to produce signals which cause valves, etc., to close, open, and so forth. A jet of oil issues from the nozzle and impinges on the flapper. Derive an expression for the force F which would be necessary to hold the flapper in equilibrium against the jet reaction. Let d_n = nozzle diameter, and P_n = pressure in nozzle.

CHAPTER 12

Some further considerations of hydrodynamics

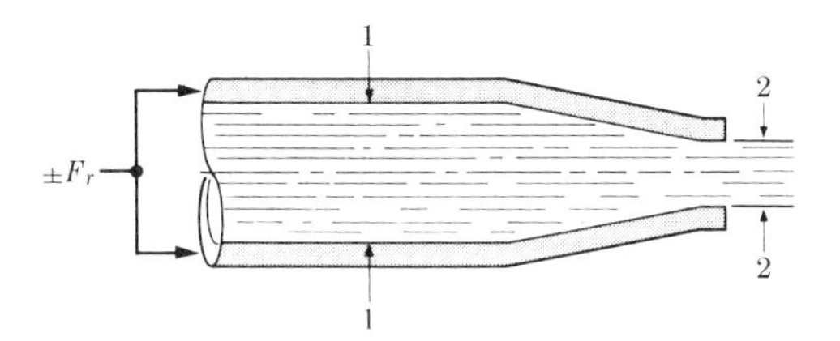

Fig. 12–1

When fluid flows in a conductor, such as a pipe, it is subject to the same fundamental laws as a free jet. Any difference in action between the two is a result of the constraint exerted by the conductor on the fluid stream. Thus, if we have a nozzle, as shown in Fig. 12–1, there will be one component of reaction on the pipe due to static pressure, and another due to the effects of change in momentum. We can calculate the net reaction on the pipe by summing forces along the centerline of the flow path:

$$\Sigma F = F_s + F_j \pm F_r = 0.$$

The static pressure force is $F_s = P \cdot A_1$. The jet reaction force is

$$F_j = M_1(v_1 - v_2)$$

as determined in the previous chapter. Thus

$$\pm F_r = P_1 \cdot A_1 + \frac{\gamma Q^2}{g}\left(\frac{1}{A_1} - \frac{1}{A_2}\right),$$

where γQ = mass flow rate, and $v = Q/A$. The reason for the "plus or minus" in front of the reaction symbol F_r is that it cannot be determined in which direction the net reaction will occur until the particular system in question is evaluated. If the pressure force is the larger of the two components, the net reaction will be to the right. Conversely, if the jet reaction force is greater, the net reaction will be to the left.

Pipe Bend

The pipe bend shown in Fig. 12–2 is subject to reaction forces due to pressure and momentum phenomena. The flow rate of the fluid in the pipe is Q. According to the principle of flow continuity, the velocity will be $v = Q/A$.

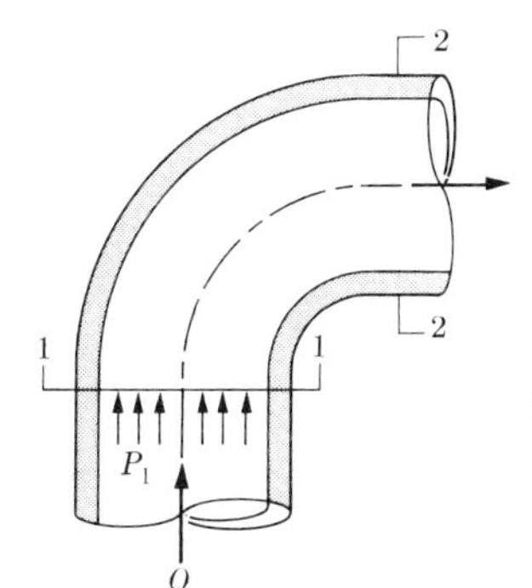

Fig. 12–2

Summing forces in the y-direction, we determine the following:

1) There is a pressure force: $F_{y_p} = P_1 \cdot A_1$.
2) The reaction force due to change in momentum is

$$F_{y_j} = M_1(v_2 - v_1),$$

where $M_1 = \gamma Q/g = \gamma v_1 A_1/g$, v_1 is calculated from Q/A_1, and $v_2 = 0$ because there is no y component beyond the bend. Thus

$$-F_{y_j} = \frac{\gamma A_1 v_1^2}{g},$$

where the minus sign means that the reaction is opposite in sense to the direction of flow. However, F_{y_j} represents the reaction of the pipe wall on the fluid, so that the reaction of the fluid on the pipe wall would have the opposite sense. Thus the sum of the forces in the y-direction is

$$F_y = P_1 A_1 + \gamma A_1 v_1^2/g.$$

Since the bend is symmetrical, the sum of the reaction forces in the x-direction is of the same magnitude. The total reaction on the pipe due to flow around the bend will be

$$F_t = \sqrt{2}\, F_y,$$

where the direction is at 45° to the centerline of the pipe and toward the "outside" of the bend. (See Fig. 12–3.)

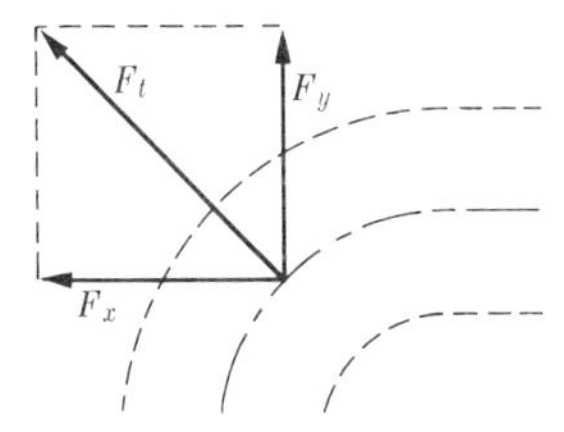

Fig. 12–3

Note that the above solution implies no losses. If we consider the pressure loss which would actually occur, then $P_1 > P_2$. Thus $F_y \neq F_x$, and each of the reactions would have to be calculated independently and combined afterward.

EXAMPLE

Given a pipe bend similar to that of Fig. 12–2 in which 80 gpm of hydraulic oil ($S_g = 0.8$) is flowing. The pipe is 2 in. in diameter. The pressure at point 1 is 100 psi, and there is a head drop of 14.5 ft between points 1 and 2. Determine the reaction of the oil on the pipe bend.

Solution.

Step 1. Evaluate the parameters and put into consistent units:

a) $80 \text{ gpm} = 80 \times 231\frac{1}{60} = 308 \text{ in}^3/\text{sec},$

b) $$v_1 = \frac{Q}{A} = \frac{308 \text{ in}^3/\text{sec}}{3.1416 \text{ in}^2} = 98 \text{ in/sec},$$

where A is the cross-sectional area for a 2-in. diameter,

c) $P_2 = P_1 - P_L = 100 - 5 = 95$ psi,

where $P_L = 0.433\, S_g h_L = 0.433 \times 0.8 \times 14.5 = 5$ psi,

d) $\gamma = 0.8 \times 0.0361 \text{ lb/in}^3 = 0.0289 \text{ lb/in}^3.$

Step 2. Calculate the pressure forces:

a) $F_{P_1} = 100 \times 3.1416 = 314.16$ lb,

b) $F_{P_2} = 95 \times 3.1416 = 298.2$ lb.

Step 3. Calculate the hydrodynamic forces:

$$F_y = M_1(v_2 - v_1) = \frac{308 \times 0.0289}{386} (0 - 98)$$

$$= 2.25 \text{ lb} = F_x,$$

since the bend is symmetrical and the pipe is of constant cross-sectional area.

Step 4. Sum forces in the x- and y-directions:

a) $F_x = 298.2 \text{ lb} + 2.25 \text{ lb} = 300.45$ lb

b) $F_y = 314.16 \text{ lb} + 2.25 \text{ lb} = 316.41$ lb

c) $F_t = \sqrt{300.45^2 + 316.41^2} = 435$ lb at an angle of 47°44′ to the horizontal.

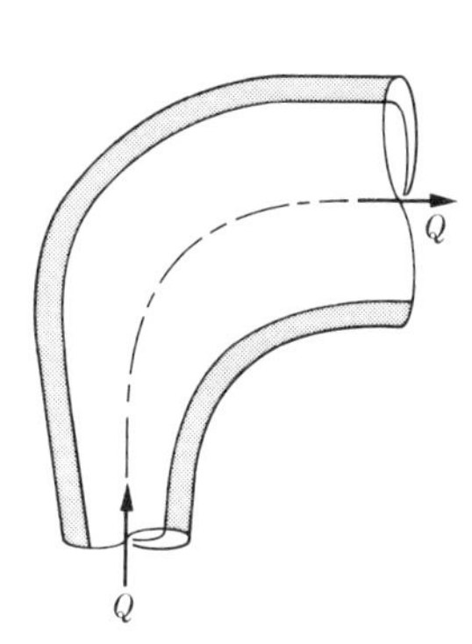

Fig. 12–4

It is apparent that if the pipe bend is not symmetrical and/or of constant cross-sectional area, the hydrodynamic forces in the x- and y-directions will not be the same. As shown in Fig. 12–4, there will be an x- and a y-component of exit velocity. There will also be a change in flow velocity due to the change in section. This follows from considerations of flow continuity: $Q = A_1v_1 = A_2v_2$. And finally, the pressure forces will not be symmetrical because of the different areas and the pressure drop through the bend.

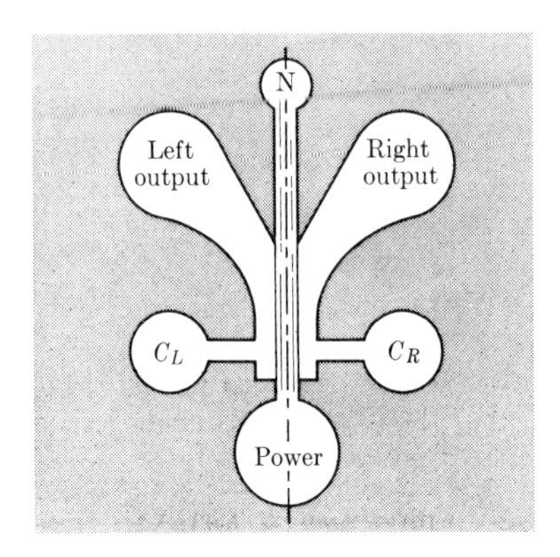

Fig. 12–5

The Pure-Fluid Amplifier*

The device illustrated in Fig. 12–5 represents a typical momentum-exchange mechanism, known as a *pure-fluid* amplifier. Based on the principles of momentum conservation just discussed, this concept has opened the door to a new area in control technology. The device consists of a main or power jet issuing from a central nozzle. Two control jets are positioned at right angles to the power jet, one on each side of the power stream. So long as no control signal (a jet of fluid) issues from either the right-hand or left-hand control jet, the main jet issues straight from the nozzle and exits through the neutral port, N. If a control jet issues from the left-hand control jet, as shown in Fig. 12–6, the main power jet will be deflected toward the right. If the right control sends forth a jet, the power jet will be deflected toward the left. The angle of deflection of the power jet

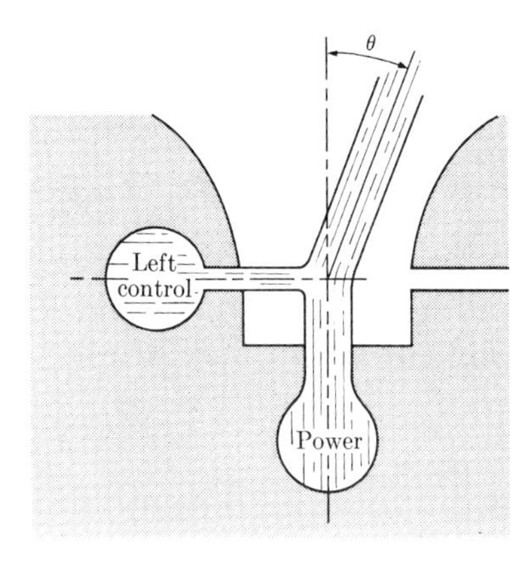

Fig. 12–6

---

* A new technology based on this and similar devices is evolving. It is called "Fluidics" and deals with logical control using fluid devices.

Fig. 12–7. Plastic models of jet-type fluid amplifiers.

is a function of the momenta of the control jets. This type of fluid amplifier is called a *proportional* amplifier, since the main jet deflection is proportional to the strength of the control signal. Figure 12–7 shows a jet interaction type of fluid amplifier molded of clear plastic.

EXAMPLE

The above discussion can be illustrated mathematically as follows:

Step 1. The momentum of the main (power) jet is $m_p v_p$.

Step 2. The momentum of one control jet is $m_c v_c$, where m = mass and v = velocity.

Step 3. Then $\tan\theta = v_t/v_a$, where

$$v_t = \frac{m_c v_c}{m_c + m_p}, \qquad v_a = \frac{m_p v_p}{m_c + m_p}.$$

Step 4. From step 3, it follows that

$$\tan\theta = \frac{m_c v_c}{m_p v_p}.$$

Step 5. Note that $mv = \rho A v v = \rho A v^2$.

Step 6. Thus assuming $\rho_c = \rho_p$, we have

$$\tan\theta = \frac{\rho_c A_c v_c^2}{\rho_p A_p v_p^2}.$$

Step 7. From Torricelli's equation: $v = C_v\sqrt{2gh}$, $v^2 = K \cdot P$.

Step 8. Thus the equation for the tangent of the angle of deflection could be written in the form

$$\tan \theta = KP_c/KP_p;$$

that is, the angle of deflection is that angle whose tangent is the ratio of the control pressure to the power pressure.

The power jet impinges on the *splitter* between the neutral port and one of the outlet ports. The passages of the outlet ports function as *diffusers*, reconverting the kinetic energy of the jet to pressure energy. The degree of pressure recovery is indicative of the efficiency of the device. It can be seen that the greater the deflection of the main jet toward an outlet port, the greater the amount of energy recovered.

The study of these pure-fluid devices is rapidly becoming a branch of fluid mechanics in its own right. This brief introduction is presented here only to alert the student to the existence of this new field.

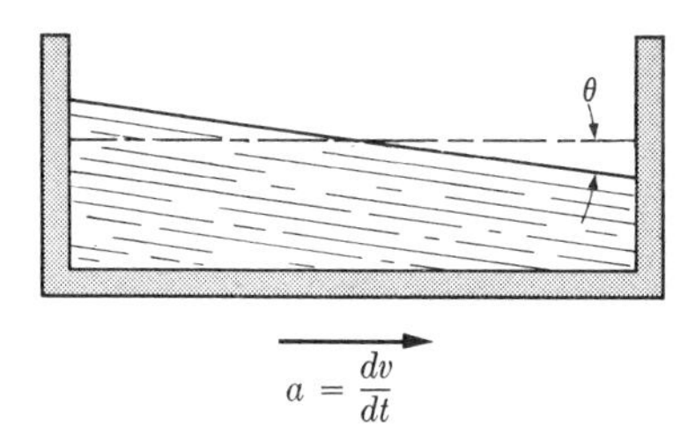

Fig. 12–8

Fluids Under Acceleration

Let us recall our definition of a fluid as being a substance which cannot sustain a shear stress under equilibrium conditions. Thus, if a tank of liquid is accelerated (see Fig. 12–8), the surface of the liquid will be "tilted" with respect to the horizontal. The angle will be

$$\theta = \tan^{-1} \frac{a}{g},$$

where a is the linear acceleration of the tank and g is the acceleration due to gravity. The pressure at any point in the fluid being accelerated vertically is given by the following expressions:

$$P_a = \gamma h(1 + a/g)$$

when the acceleration is upward, and

$$P_a = \gamma h(1 - a/g)$$

when the acceleration is downward.

EXAMPLE

We have a rectangular tank, 3 ft × 10 ft × 4 ft (see Fig. 12–9). It is filled with water to a depth of 2 ft. The tank is accelerated horizontally at 6 ft/sec^2. Compute the total force acting on each end of the tank and show that the difference between these two forces equals the unbalanced force needed to accelerate the mass of the liquid.

Solution.

Step 1. $\tan\theta = a/g = 6/32.2 = 0.1862;\ \theta = 10°33'$.

Step 2. The depth of the water at the right end is equal to the original depth (2 ft) minus the drop in head due to acceleration:

$$a - h = 2\text{ ft} - (5\tan\theta) = 2\text{ ft} - (5 \times 0.1862)$$
$$= 2 - 0.93 = 1.07\text{ ft}.$$

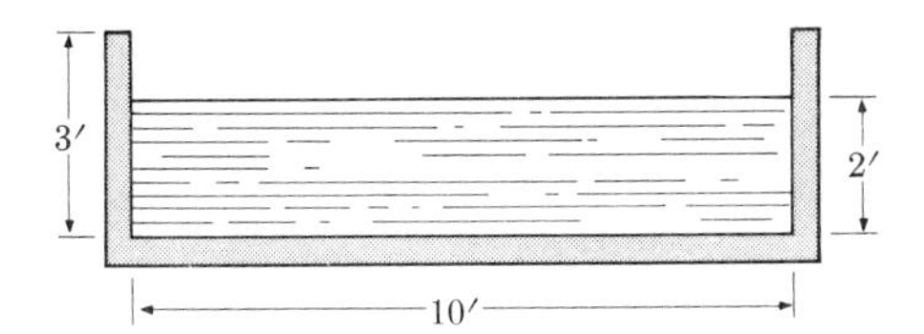

Fig. 12–9

Step 3. The depth of the water at the left end is equal to the original depth plus the rise in head due to the acceleration:

$$a + H = 2\text{ ft} + (5\tan\theta) = 2\text{ ft} + (5 \times 0.1862) = 2.93\text{ ft}.$$

Step 4. The total force on the left end (based on $F = P \times A$) is

$$a - F_H = (\gamma H/2)(H \cdot 4) = 62.4 \times (2.93/2) \times 2.93 \times 4 = 1072\text{ lb}.$$

Step 5. The total force on the right end is

$$a - F_h = (\gamma h/2)(h \cdot 4) = 62.4 \times (1.07/2) \times 1.07 \times 4 = 144\text{ lb}.$$

Step 6. The force difference equals the unbalanced force necessary to accelerate the mass of liquid:

a) $F_H - F_h = M \cdot a, \qquad 1072\text{ lb} - 144\text{ lb} = 928\text{ lb}.$

b) $\dfrac{4 \times 10 \times 2 \times 62.4}{32.2} \times 6 = 930$ lb,

which checks within slide rule accuracy.

Rotating Body of Fluid

The shape of the surface of a free fluid in a rotating reservoir is that of a paraboloid of revolution, as shown in Fig. 12–10. The equation of the resultant paraboloid is

$$y = \frac{\omega^2}{2g} x^2.$$

This formula can be verified as follows:

1) On any point mass A in the surface of the fluid at a distance x from the axis of rotation, there are three forces acting:

a) The weight of the element, W,

b) The inertia force, $(W/g)\omega^2 x$, which acts radially away from the axis of rotation, and

c) A resultant force F due to the action of surrounding fluid particles, which acts normal to the surface at A.

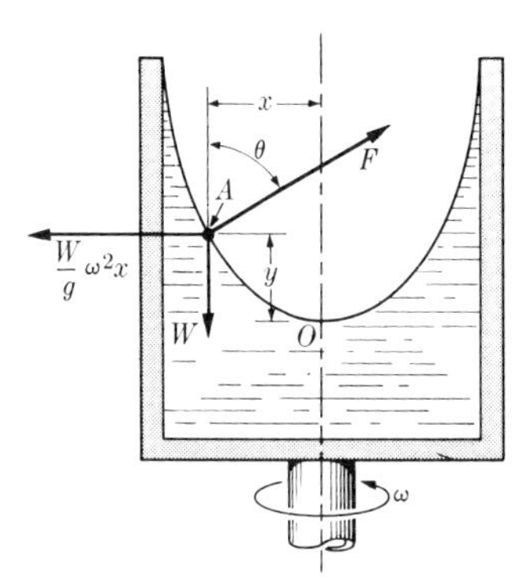

Fig. 12–10

2) These three forces are in equilibrium. Thus

a) $F \sin \theta = (W/g)\omega^2 x,$

b) $F \cos \theta = W.$

3) Dividing (a) by (b), we have

$$\tan \theta = x\omega^2/g.$$

4) Note that $\tan\theta$ is also the slope of the curve at point A:

$$A = dy/dx.$$

5) Thus

$$\frac{dy}{dx} = \frac{\omega^2 x}{g} \quad \text{and} \quad y = \int \frac{\omega^2 x}{g}\,dx = \frac{\omega^2 x^2}{2g} + C_1.$$

6) When $x = 0$ and $y = 0$, then $C_1 = 0$. Thus $y = \omega^2 x^2/2g$.

The pressure in a closed reservoir will be increased by rotation of the reservoir. The increase in pressure between a point on the axis of rotation and a point x units away from the axis is given by

$$P = \gamma \frac{\omega^2 x^2}{2g}.$$

Similarly, the increase in pressure head is

$$P/\gamma = \omega^2 x^2/2g = y.$$

Note that this is the same as the expression developed above.

IMPORTANT TERMS

Reaction force of a fluid on a conductor is the force developed due to the change in momentum of the fluid stream.

Pure-fluid amplifier is a new class of fluid devices in which relatively large amounts of energy are controlled by small amounts.

PROBLEMS

12–1 Consider a nozzle similar to that shown in Fig. 12–1. Let the area at 1–1 be 3.1416 in^2 and that at 2–2 be 0.7854 in^2. The pressure at 1–1 is 75 psi. The fluid is water. Its flow rate is 50 gpm. Determine the magnitude and direction of the restraining force required to hold the nozzle in equilibrium.

12–2 A nozzle with an exit diameter of 1 in. is located at the end of a 2-in. (diameter) pipe. The pressure in the pipe is 50 psi. If it takes force of 500 lb to restrain the pipe, what is the flow rate of MIL-5606 through the pipe?

12–3 A nozzle with an exit diameter of $1\frac{1}{2}$ in. is located at the end of a 4-in. pipe. If 400 gpm of sea water flow through the system and the restraining force required to hold the pipe in equilibrium is 600 lb, what is the pressure in the pipe?

12–4 An orifice 2 in. in diameter is located in the end of a 5-in. pipe, as shown in Fig. 12–11. The pressure in the pipe is 100 psi. The fluid is glycerine. The flow rate is 350 gpm. Calculate the restraining force required to maintain the pipe in equilibrium.

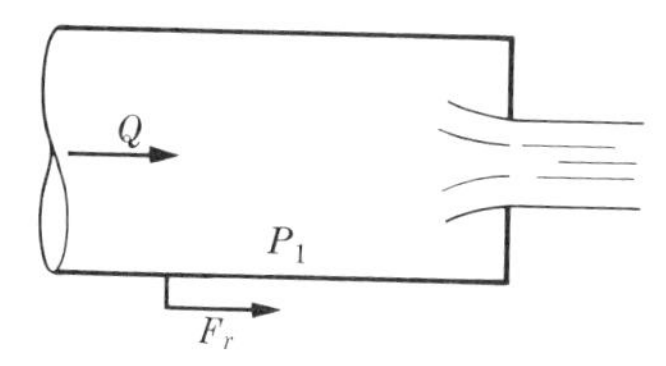

Fig. 12–11

12–5 One arm of a rotating lawn sprinkler is shown in Fig. 12–12. Each nozzle is $\frac{3}{4}$ in. in diameter and the flow rate of the water passing through the sprinkler is 20 gpm. The tangential velocity at the tip of the sprinkler arms is 6 ft/sec. Determine the net torque due to the water flow through the two arms of the sprinkler.

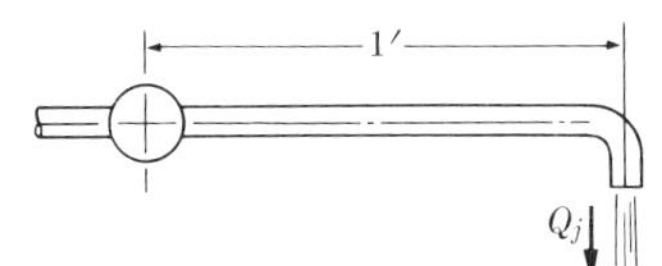

Fig. 12–12

12–6 A jet flows through an orifice in the side of a tank. Derive the expression for the reaction produced by the jet on the tank.

12–7 Assuming no friction ($C_v = 1$), prove that $C_c = 0.5$ for the Borda mouthpiece discussed in Chapter 9.

12–8 The illustration in Fig. 12–13 shows a jet pipe, a fluid control device used in industry to produce output pressures proportional to a control signal. The jet pipe pivots about a vertical centerline through a small angle $\pm\theta$. As it pivots, the jet of fluid issuing from the nozzle at the end of the pipe is split between the receiver ports in proportion to the angle θ. In the center position the jet splits evenly and both ports 1 and 2 develop the same output pressure. Assume that the jet pipe has a diameter of $\frac{1}{4}$ in. and that the nozzle at its end has a diameter of $\frac{1}{8}$ in. The supply pressure to the jet pipe is 500 psi. The fluid is hydraulic oil, and the flow rate is 11 gpm.

Calculate the reaction on the pivot due to the jet. Determine the theoretical pressure signal in each receiver port when the jet is centered.

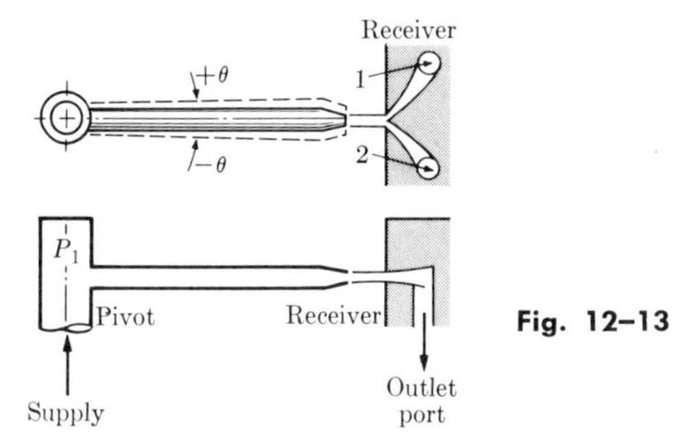

Fig. 12–13

12–9 Assume that the jet pipe of Problem 12–8 is rotated through an angle such that the jet splits three parts to port 1 and one part to port 2. The ports are connected to a hydraulic cylinder with a blank end area of 10 in^2 and a rod area of 2 in^2. What is the maximum force the cylinder can exert if port 1 is connected to the blank end port?

12–10 A conductor of varying cross section is shown in Fig. 12–14. Demonstrate that the net reaction force F_R is expressed by the following equation:

$$F_R = A_2(P_2 + \rho_2\bar{v}_2^2) - A_1(P_1 + \rho_1\bar{v}_1^2).$$

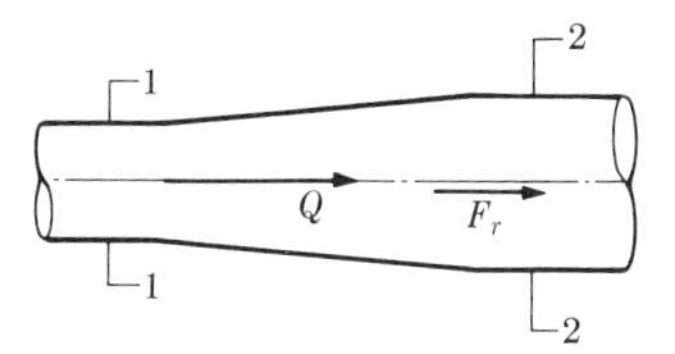

Fig. 12–14

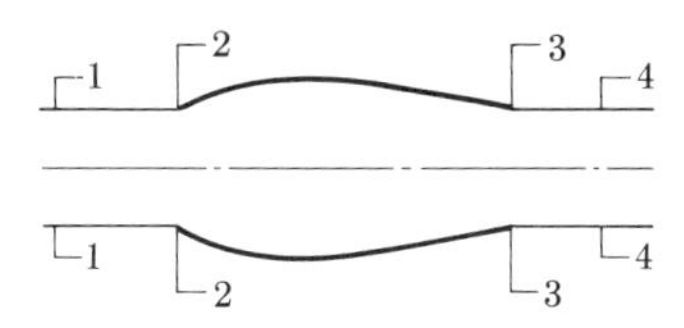

Fig. 12–15

12–11 A typical air-breathing jet engine is shown in Fig. 12–15. At section 1–1, the velocity is flight velocity. The velocity at section 4–4 is exit gas velocity. Prove that the equation for thrust is

$$F_t = \frac{W_{\text{exit}}v_4}{g} - \frac{W_1v_1}{g}.$$

12–12 The engine of Problem 12–11 is being *bench*-tested in the laboratory. The engine consumes air at the rate of 75 lb/sec and burns 1.0 lb/sec of fuel. The exit gas velocity is 2000 fps. Calculate the static thrust of the engine if entrance velocity is 1000 fps.

12–13 If the engine of Problem 12–12 is operated at 600 mph in flight, what is the exit gas velocity required to generate the same thrust as in Problem 12–12?

12–14 Consider a 90° pipe bend similar to that in Fig. 12–2. The pipe is 2 in. in diameter, the flow rate of water is 250 gpm. The pressure at section 1–1 is 150 psi. Calculate the reaction on the pipe due to the change in direction of the fluid.

12–15 A 90° pipe bend has the changing-section configuration shown in Fig. 12–4. The small section at the inlet is 3 in. in diameter, which smoothly transforms to a 6-in. diameter at the outlet section. The fluid is gasoline, and the flow rate is 400 gpm at an inlet pressure of 100 psi. Calculate the magnitude and direction of the reaction on the pipe due to the change in the direction of the fluid.

12–16 Assume that we have a pipe bend similar in specifications to that of Problem 12–15 except that the bend angle is 60° instead of 90°. Calculate the reaction under these conditions.

12–17 A rectangular tank (see Fig. 12–8) is 6 ft high, 10 ft long, and 4 ft wide. It contains water to a depth of 4 ft. If it is accelerated horizontally at 9 ft/sec^2, what is the depth of water at each end and the total force on each end?

12–18 A rectangular open tank is carried on a truck. The tank is 3 ft high, 5 ft wide, and 12 ft long. It carries 8.97 gal of hydraulic oil. What is the maximum acceleration to which the truck can be subjected without spilling any oil?

12–19 A can 10 in. in diameter and 12 in. high is filled with turpentine. It is placed in an elevator which accelerates upward at 15 ft/sec^2. What is the pressure on the bottom of the can?

12–20 A cylindrical tank 18 in. in diameter and 24 in. high is half-filled with phosphate ester hydraulic fluid. If it is rotated about its vertical axis, at what speed will the fluid reach the top of the tank?

12–21 In the system of Problem 12–20, what will be the maximum pressure?

12–22 An open cylindrical tank 5 ft high and 3 ft in diameter is filled with water to a height of 3 ft. At what constant angular velocity can it be rotated without spilling any water?

12–23 A closed cylindrical tank 4 ft high and 2 ft in diameter is filled with glycerine to a height of 3 ft. It is rotated at a speed of 145 rpm. How much of the bottom of the tank will be uncovered?

12–24 A pipe 2 in. in diameter and 6 ft long is filled with oil (S_g = 0.8) and rotated about an axis perpendicular to the centerline of the pipe at a speed of 150 rpm (see Fig. 12–16). Calculate the pressure on the cap at the end of the pipe.

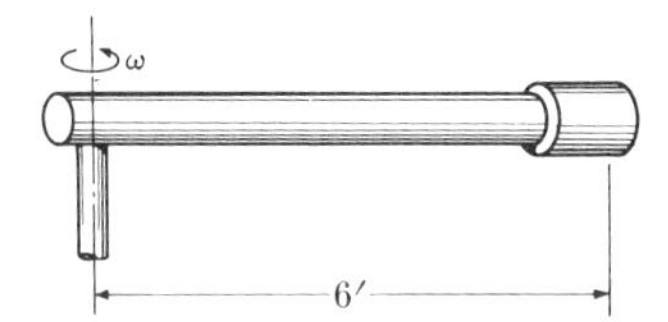

Fig. 12–16

CHAPTER 13

Flow of fluid in pipes

The material covered so far in our discussions has been based, for the most part, on the so-called *ideal fluid.* Where losses, such as head loss, etc., had to be considered, we introduced them as additonal factors in an equation, for example, H_L in Bernoulli's equation. This loss factor made the *ideal* situation conform with reality. We can justify our approach on the grounds that fluid mechanics is to a large extent an empirical science.

The topic of pipe flow, which we are about to discuss, is perhaps one of *the* most empirical areas of fluid mechanics.

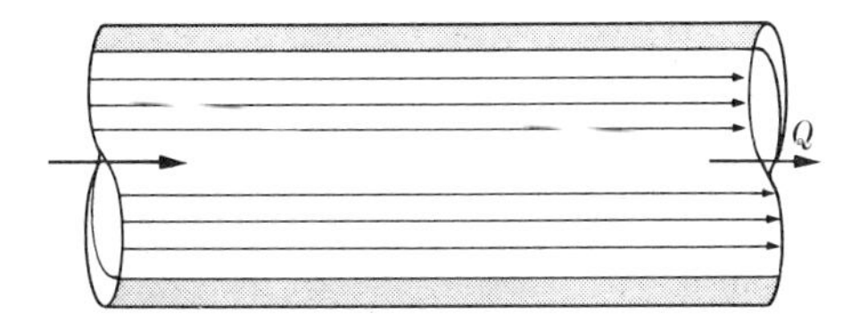

Fig. 13–1

Types of Flow

When fluids flow in conductors, they can flow in one of two modes: (1) laminar flow, (2) turbulent flow. In laminar flow, the fluid moves in parallel layers, or *lamina,* as illustrated in Fig. 13–1. The flow streams are parallel as the fluid moves through the conductor. In turbulent flow, on the other hand, there are irregular motions. Velocity fluctuations are superimposed

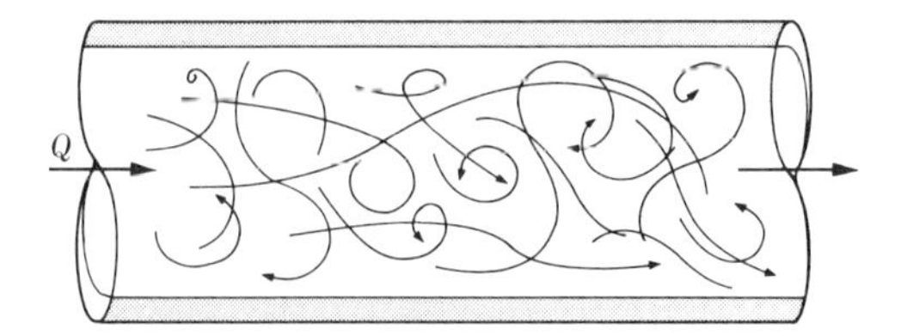

Fig. 13–2

on the main or average flow, as shown in Fig. 13–2. This swirling flow results in increased losses as energy is dissipated by the interaction of the fluid. An interesting way of demonstrating turbulent flow is to observe the smoke column given off by a cigarette or snuffed-out candle. The smoke will rise in a smooth even column for some distance. The column will then break up into a swirling, eddying motion: laminar has changed to turbulent flow.

Critical Velocity

The term *critical velocity* is applied to the velocity at which the transition from laminar to turbulent flow occurs. This transition cannot be pinpointed precisely, as can be seen in Fig. 13–3. Let us suppose that we observe laminar flow under a given set of conditions. As the flow velocity increases, as shown in the figure, it reaches a critical value above which the flow will be turbulent. This is called the *upper critical velocity*. Again, let us suppose that we observe turbulent flow. The flow velocity decreases continuously until it reaches some value below which the flow changes to laminar flow. This is termed the *lower critical velocity*.

The region between the lower and upper critical velocities is a transition region. The flow conditions and the side of critical velocity (upper or lower) from which we are approaching determine whether we will encounter laminar or turbulent flow in this region.

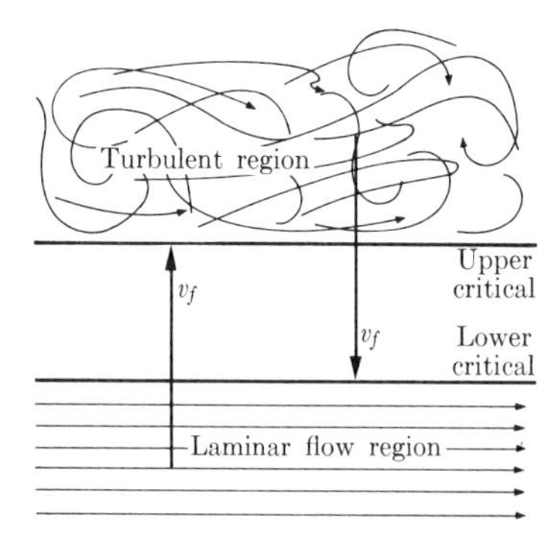

Fig. 13–3

Note that this phenomenon reflects the tendency of all natural systems to remain just as long as possible in the state in which they currently exist. Laminar flow tries to remain laminar until it reaches velocity levels which preclude this possibility; the same holds true for turbulent flow.

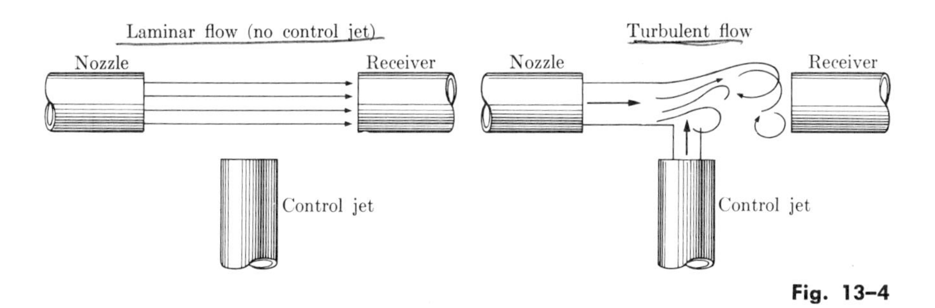

Fig. 13–4

Flow-Mode Amplifier

In Chapter 12, we discussed the pure-fluid amplifier, a device based on the momentum exchange between two interacting jets of fluids. There is another class of fluid-control devices based on the phenomenon we have just discussed, that is, the change of the flow mode from laminar to turbulent under certain conditions. This flow-mode device, illustrated in Fig. 13–4, is called a *turbulence amplifier*. When the velocity of the jet issuing from the nozzle is below the threshold value required for laminar flow, we have a uniform stream between the nozzle and the receiver. If the velocity is too high or, if a disturbance is introduced (the method used in the device), the flow will switch to turbulent. Figure 13–5 shows a pressure-recovery curve for a typical amplifier. Under laminar flow conditions the pressure recovery is a maximum. As the turbulence increases, recovery lessens until full turbulent flow results in minimum values of pressure at the receiver. In utilizing this device in fluid logic applications, the higher

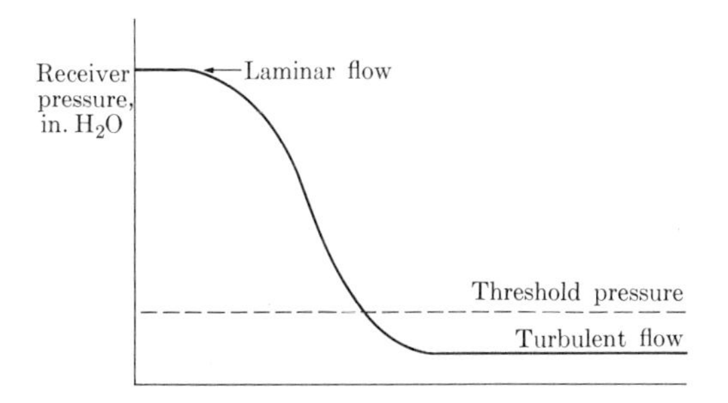

Fig. 13–5

pressure can be considered to be an "on" signal, and the very low pressure an "off" signal. A threshold pressure, below which the system with the amplifier will not work, is set above (dashed line) this lower pressure. Thus the device will provide true on-off switching.

Reynolds Number

Because of the large number of variables encountered in pipe flow systems, it is impractical to attempt an analytical solution. Current practice uses empirical methods to solve pipe flow problems.

Among the more important empirical criteria established for pipe flow is the Reynolds number. This is the criterion which tells us whether we will experience laminar or turbulent flow.

$$N_R = \frac{D \cdot v \cdot \rho}{\mu} = \frac{Dv}{\nu},$$

where D = pipe diameter,
v = velocity,
ρ = fluid density,
μ = absolute viscosity,
ν = kinematic viscosity.

Since the Reynolds number is a dimensionless ratio, all units must be consistent in the above expression. The Reynolds numbers relates the basic flow factors: pipe diameter, average velocity of flow in the pipe, and the viscosity of the fluid.

For pipe flow, it has been determined that average Reynolds-number values of less than 2000 indicate laminar flow. This corresponds to the lower critical velocity. When the Reynolds number is above 4000, the flow will be turbulent. This would correspond to upper critical velocity. Since so many factors such as entrance conditions, pipe roughness, initial disturbances, etc., determine the actual state of fluid flow, these values of Reynolds numbers are only approximate.

This discussion of the Reynolds number has introduced a new term, viscosity, which was not encountered heretofore.

Viscosity

At the outset of the study of fluid mechanics, we defined a fluid as a substance which could not sustain a shear stress under equilibrium conditions. We also agreed to approach the early part of our study of fluid mechanics from the point of view of an ideal fluid, in which we ignored losses. We subsequently included losses in a flowing system as a correction factor, H_L. It is here, in the concept of *viscosity*, that we face head-on the idea of losses in fluid flow.

> **Viscosity** is that property of a fluid by which it offers resistance to shear stresses.

This is not inconsistent with our definition of fluid, since the conditions under which we encounter viscous forces, or shearing forces, are not equilibrium conditions.

In developing a picture of the nature of viscosity, we must consider three basic premises:

1) Fluid in contact with a surface has the same velocity as that surface.

2) As we move away from the surface, the rate of change of velocity is uniform.

3) The shearing stress in the fluid is proportional to the rate of change of velocity.

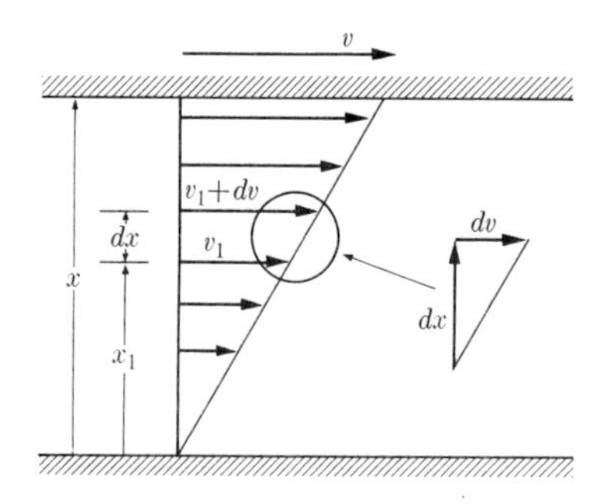

Fig. 13–6

These assumptions are illustrated in Fig. 13–6, in which we show two parallel plates separated by a finite distance. The lower plate is stationary; thus its velocity is zero. The upper plate moves relative to the lower with some velocity v. According to the first premise, the velocity of the layer of fluid in contact with the lower plate is zero, while the velocity of the layer of fluid in contact with the top plate is v. By the second premise, the velocity distribution between the plates is uniform, that is, a straight line. This is illustrated by the triangle in Fig. 13–6: the base of the triangle is the vector representing the velocity of the upper plate, v, and the apex of the triangle lies at the fixed lower plate where the velocity is zero.

If we take a strip of fluid of incremental width dx at any random position between the plates, x_1, then the velocity at the lower border of the strip will be v_1. That at the upper boundary of the strip will be $v_1 + \Delta v$.

According to the second assumption, the velocity change will be uniform. Thus the change in velocity for a change of distance, dx, is dv. By similar triangles, $v/x = dv/dx$. The third premise states that the shear stress is proportional to the rate of change of velocity, dv/dx. Thus

$$\tau = \mu \, dv/dx,$$

where τ is shear stress in units of force per unit area and μ is the constant of proportionality. Since $v/x = dv/dx$, $\tau = \mu v/x$, and $\mu = \tau x/v$.

If the system is adjusted so that the plates are a unit distance apart, $x = 1$, and if the upper plate has unit velocity, $v = 1$, then $\tau = \mu$. In this case, μ is called the *absolute or dynamic viscosity*.

Absolute viscosity is defined as the force required to move a flat plate of unit area at unit distance from a fixed plate with unit relative velocity when the space between the plates is filled with the fluid whose viscosity is being measured.

Units of absolute viscosity are as follows:

1) In the foot-pound-second system:

$$\tau = \text{lb/ft}^2, \qquad x = \text{ft}, \qquad v = \text{ft/sec};$$

$$\mu = \frac{\tau x}{v} = \frac{(\text{lb/ft}^2) \times \text{ft}}{\text{ft/sec}} = \frac{\text{lb} \cdot \text{sec}}{\text{ft}^2}.$$

2) In the metric system:

$$\tau = \text{dyne/cm}^2, \qquad x = \text{cm}, \qquad v = \text{cm/sec};$$

$$\mu = \frac{\tau x}{v} = \frac{(\text{dyne/cm}^2) \times \text{cm}}{\text{cm/sec}} = \frac{\text{dyne} \cdot \text{sec}}{\text{cm}^2} = \text{poise}.$$

The poise turns out to be a relatively large unit; hence the centipoise is generally used: 1 centipoise = 1 poise/100.

3) At 68°F (20°C), the experimentally determined value of absolute viscosity of water is one centipoise.

4) *Relative viscosity* is the ratio of the absolute viscosity of any fluid to that of water at 68°F.

Kinematic Viscosity

The ratio of the absolute viscosity of a fluid to its mass density is called the *kinematic viscosity*. The symbol ν is generally used to represent kinematic viscosity: $\nu = \mu/\rho$.

Units of kinematic viscosity are as follows:

1) In the foot-pound-second system:

$$\nu = \frac{\mu}{\rho} = \frac{\text{lb} \cdot \text{sec/ft}^2}{\text{lb} \cdot \text{sec}^2/\text{ft}^4} = \frac{\text{ft}^2}{\text{sec}}.$$

2) In the metric system:

$$\nu = \frac{\mu}{\rho} = \frac{\text{dyne} \cdot \text{sec/cm}^2}{\text{dyne} \cdot \text{sec}^2/\text{cm}^4} = \frac{\text{cm}^2}{\text{sec}} = \text{stoke},$$

$$1 \text{ centistoke} = \frac{1 \text{ stoke}}{100}.$$

Conversion from English to Metric Systems of Measure

One of the aggravating aspects of fluid mechanics, at least here in the United States, is the necessity of handling the units of the English system of measure. When we must convert units from the English to the metric system, the problems become even more pronounced. Since most sources of viscosity data give the values in metric units, the following conversion data will prove useful.

1) Absolute viscosity:

$$\mu = \frac{\text{lb} \cdot \text{sec}}{\text{ft}^2} = \frac{453.6 \text{ gm/lb} \times 980.665 \text{ dynes/gm}}{(30.48 \text{ cm/ft})^2}$$

$$= 478.8 \text{ dyne} \cdot \text{sec/cm}^2 \text{ (poise)} = 47{,}880 \text{ centipoise},$$

$$\mu = \frac{\text{lb} \cdot \text{sec}}{\text{in}^2} = \frac{453.6 \text{ gm/lb} \times 980.665 \text{ dynes/gm}}{(2.54 \text{ cm/in})^2}$$

$$= 68{,}948 \text{ poise} = 6.8948 \times 10^6 \text{ centipoise};$$

2) Kinematic viscosity:

$$\nu = \frac{\text{ft}^2}{\text{sec}} = \frac{(30.48 \text{ cm/ft})^2}{\text{sec}} = 929.03 \frac{\text{cm}^2}{\text{sec}} \text{ (stoke)}$$

$$= 92{,}903 \text{ centistokes}.$$

EXAMPLE

Two parallel plates are positioned $\frac{1}{2}$ in. apart in a manner similar to that shown in Fig. 13–6. The space is filled with an oil which has an absolute viscosity of $3160 \times 10^{-5}(\text{lb} \cdot \text{sec})/\text{ft}$. If the upper plate moves with a velocity of 8 ft/sec and the lower one is stationary, what is the shear stress in the oil?

$$\tau = \mu \cdot \frac{v}{x} = 3160 \times 10^{-5} \frac{8}{0.5/12}$$
$$= 6.06 \text{ lb/ft}^2.$$

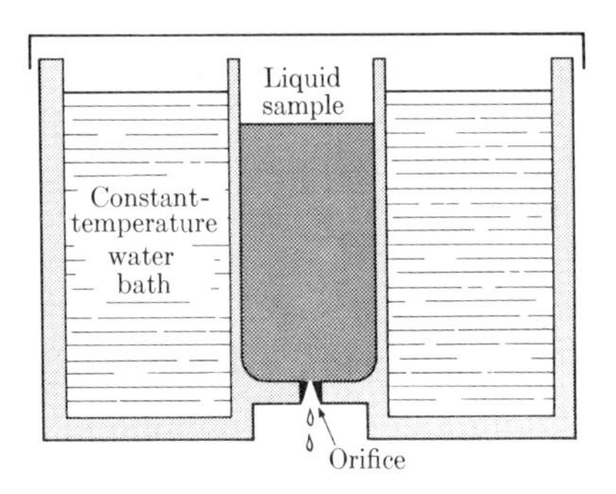

Fig. 13–7

Viscosity: Saybolt Seconds Universal

Because the determination of absolute viscosity is a matter for the laboratory, the American Society for Testing Materials has adopted the Saybolt viscosimeter for quick determinations of the viscosity of fluids. A schematic drawing of this device is shown in Fig. 13–7. Basically it consists of a temperature-controlled bath into which is placed a sample of the fluid to be tested. A standard orifice is positioned in the bottom of the apparatus. A sample of 60 cc of the liquid is collected, and the time required for it to flow through the metering orifice into the container is measured. This time

TABLE 13–1

SAE No.	*Viscosity range, SSU*			
	at 0°F		*at* 210°F	
	Minimum	*Maximum*	*Minimum*	*Maximum*
5W	–	$<$4,000	–	–
10W	6,000	12,000	–	–
20W	12,000	48,000	–	–
20	–	–	45	Less than 58
30	–	–	58	Less than 70
40	–	–	70	Less than 85
50	–	–	85	Less than 110

is then the measure of the viscosity in Saybolt Seconds Universal, SSU. The relationship between viscosity and kinematic viscosity is shown below.

1) When (SSU) $t < 100$ sec: $\nu = 0.226t - 195/t$ centistokes.

2) When $t > 100$ sec: $\nu = 0.220t - 135/t$ centistokes.

The viscosity numbers tabulated by the Society of Automotive Engineers (SAE) are related to viscosities given in SSU. SAE viscosity numbers with a suffix "W" are based on oil viscosities at 0°F. SAE numbers without the "W" are based on viscosities at 210°F. Table 13–1 summarizes the SAE-recommended practice.

Viscosity Index

The change in viscosity with change in temperature is an important characteristic of lubricating oils, hydraulic oils, etc. The *viscosity index* is an empirical criterion which represents this fluid characteristic, critical to equipment which must operate in widely varying temperature environments, like the machine shown in Fig. 13–8. Oils of *low viscosity index* exhibit a *large change* in viscosity with temperature change. Oils of *high viscosity index* exhibit a *small change* in viscosity with change in temperature.

Fig. 13–8

Typical of a broad market for small construction equipment is the all-hydraulic backhoe shown in the illustration above. This unit demonstrates, as well as any, one of the major advantages from the geometry of the vehicle or machine. Can you conceive of transmitting the power to that bucket on the end of the boom by some other method?

The formula for calculating the viscosity index is

$$VI = \frac{L - U}{L - H} \times 100,$$

where L = viscosity at 100°F of an oil of zero VI (a reference oil), which has the same viscosity at 210°F as the sample oil;

U = viscosity of the sample oil at 100°F; and

H = viscosity at 100°F of an oil of 100 VI (a reference oil), which has the same viscosity at 210°F as the sample oil.

IMPORTANT TERMS

Laminar flow is fluid flow in which motion proceeds in parallel layers, or lamina.

Turbulent flow is fluid flow which does not proceed in parallel layers, but exhibits an irregular intermixing motion.

Critical velocity is that flow velocity at which the mode of fluid flow changes from laminar to turbulent, or vice versa.

Reynolds number is a dimensionless ratio which is the criterion determining the mode of flow that will prevail:

$$N_R = \frac{D \cdot v \cdot \rho}{\mu} = \frac{D \cdot v}{\nu}.$$

Viscosity is that property of a fluid by which it resists a shear stress.

Absolute viscosity (μ) is the force required to move a flat plate of unit area at unit distance from a fixed plate with unit relative velocity when the space between the plates is filled with the fluid.

Kinematic viscosity (ν) is the ratio of the absolute viscosity to the mass density of the fluid.

SSU, Saybolt Seconds Universal, is an empirical measure of the viscosity of a fluid.

Viscosity index is an empirical measure indicating the change in viscosity with change in temperature.

PROBLEMS

13–1 What is the kinematic viscosity, in fps units, of hydraulic oil at 70°F? 90°F? 120°F?

13–2 Determine the kinematic viscosity, in metric units, under the conditions of Problem 13–1.

13–3 The kinematic viscosity of a fluid is 0.3 ft^2/sec and the specific gravity is 0.9. Determine the absolute (or dynamic) viscosity.

13–4 A liquid ($S_g = 1.02$) has a dynamic viscosity of 1.70 centipoise. Determine the dynamic and kinematic viscosities in foot-pound-second units. What is the kinematic viscosity in stokes?

13–5 The viscosity of water at 68°F is 1.008×10^{-2} poise. What is the absolute viscosity in foot-pound-second units? Water at 68°F has $S_g = 0.998$. Calculate the kinematic viscosity in foot-pound-second units. What is ν in the metric system?

13–6 Convert a viscosity of 480 SSU to kinematic viscosity in foot-pound-second units.

13–7 Convert a viscosity of 200 SSU to kinematic viscosity.

13–8 Convert a viscosity of 75 SSU to kinematic viscosity.

13–9 Sea water flows at 10 gpm in a $\frac{3}{4}$-in. pipe. Calculate the Reynolds number.

13–10 Phosphate ester hydraulic fluid flows in a 1-in. pipe at 40 gpm. Determine the Reynolds number. Is the flow laminar or turbulent?

13–11 The flow rate of a water-glycol hydraulic fluid is 50 gpm in a 1-in. (diameter) pipe. At 90°F, will the flow be laminar or turbulent?

13–12 Using a 1-in. pipe, calculate and compare the maximum flow rates for laminar flow conditions, for each of the basic types of hydraulic fluids listed in the Appendix.

13–13 What would be the significance of designing for laminar or turbulent flow in a fluid power system?

13–14 Two flat plates are parallel to each other as shown in Fig. 13–5. The plates are $\frac{1}{4}$ in. apart; the top one moves with a velocity of 6 ft/sec, while the bottom one is stationary. If the space between them is filled with hydraulic oil at 80°F, what is the shear stress in the oil?

13–15 Two flat parallel plates are $\frac{1}{2}$ in. apart. They move relative to each other with a velocity of 10 fps. If the space between them is filled with phosphate ester, what is the shear stress in the fluid?

13–16 A vertical cylinder 4 in. in diameter rotates within a fixed sleeve or tube having an inside diameter of 4.002 in. The uniform annular space between the cylinder and sleeve is filled with hydraulic oil at 60°F. What is the resistance to rotation if the cylinder is 10 in. long and rotates at 30 rpm?

13–17 For Problem 13–16, determine the resistance to rotation, given that the fluid is gasoline instead of hydraulic oil.

13–18 It is desired to select an oil of a viscosity index suitable for use in an off-the-highway truck which will operate in summer and winter. What kind of VI characteristic would you specify and why?

CHAPTER 14

Flow of fluids in pipes (continued)

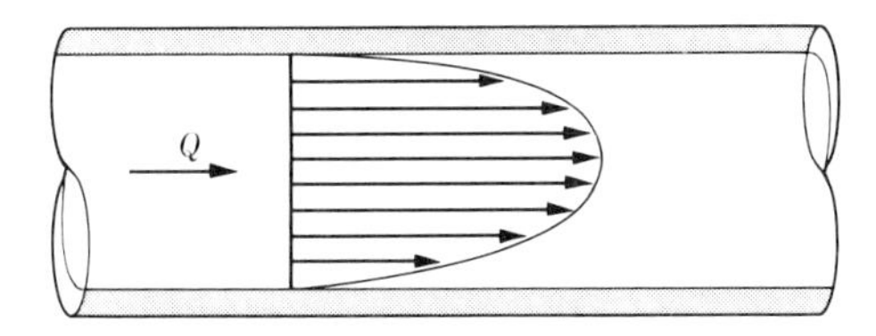

Fig. 14–1

Velocity Distribution in Pipe Flow

In the development of the concept of viscosity in the preceding chapter, one of the criteria was that the layer, or stratum, of fluid in contact with the surface had the velocity of the surface. This seems to indicate that when fluid flows between the walls of a conductor there are layers which have zero velocity.

On the other hand, when we discussed the concept of flow continuity, $Q = A_1v_1 = A_2v_2$, velocity appeared as a unique quantity, v fps or ips.

The two concepts seem incompatible. Actually there is a velocity distribution across any conductor of fluid. The fluid velocity at the wall *is* zero. The velocity varies in a parabolic manner, as shown in Fig. 14–1, with maximum velocity occurring at the center of the conductor. The velocity used in the flow-continuity relationship is really the *average* velocity in the conductor. If laminar flow takes place between two parallel plates, the average velocity is two-thirds the maximum velocity occurring at the center, that is

$$\bar{v} = \tfrac{2}{3}v_m.$$

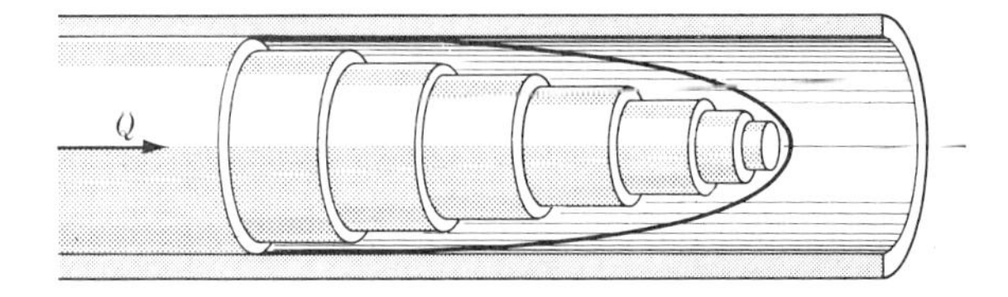

Fig. 14–2

If the fluid conductor is a pipe of circular cross section, the average velocity is one half the maximum velocity; under laminar flow conditions,

$$\bar{v} = \tfrac{1}{2}v_m.$$

A useful method for visualizing fluid flow in a pipe is illustrated in Fig. 14–2. If flow is laminar, each fluid stratum is moving parallel to the others. Each stratum, or layer, is moving with a slightly different velocity from those adjoining it. We can visualize this fluid motion as a series of telescoping tubes, each one sliding within the other with a parabolic velocity distribution.

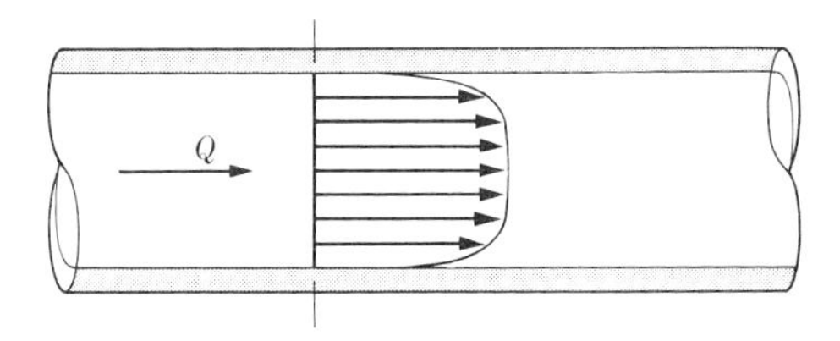

Fig. 14–3

If the flow is turbulent, the parabolic velocity distribution is destroyed. Turbulence tends to flatten out the velocity distribution curve (Fig. 14–3). The average velocity can no longer be determined analytically, as it was in the laminar case.

The velocity distribution in a curved pipe will be modified by the radius of curvature and its variation across the pipe section. Because we must conserve momentum and because the tangential velocity is a function of the radius, the distribution will look like that shown in Fig. 14–4. A secondary flow, called circulation, will be set up, as shown by the large arrow in the figure. A full discussion of the phenomena is beyond the scope of this book.

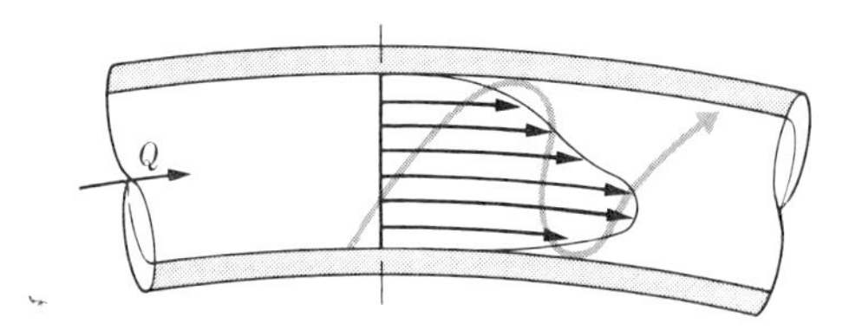

Fig. 14–4

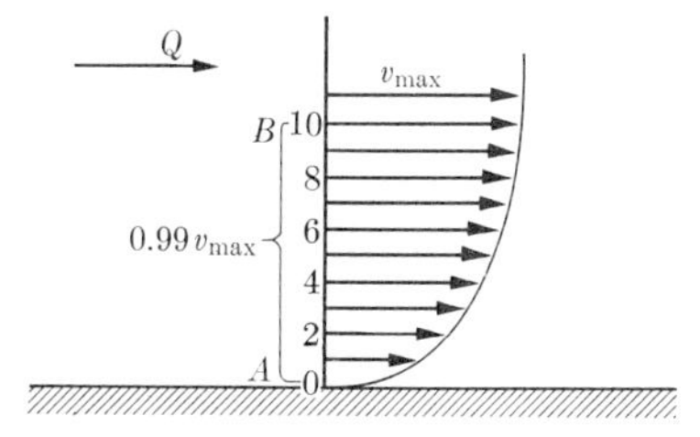

Fig. 14–5

Boundary Layer

When we talked about the criteria for defining viscosity, we considered the fact that the layer of fluid in contact with a surface has the same velocity as that surface. We noticed in Fig. 14–1 how this affected the velocity profile in a fluid conductor. The relatively thin layer of fluid in contact with a surface is called the *boundary layer*, which is illustrated in Fig. 14–5. The boundary layer is considered to be the layer of fluid adjacent to the wall of the conductor in which the velocity is 99% or less of the maximum velocity of the fluid. This region is shown from A to B in Fig. 14–5. Note that velocity vectors are numbered 1 through 10. At 0 the fluid velocity is zero; at 10 the velocity would be equal to maximum. Between 0 and 10 the distribution is parabolic, and a velocity gradient exists.

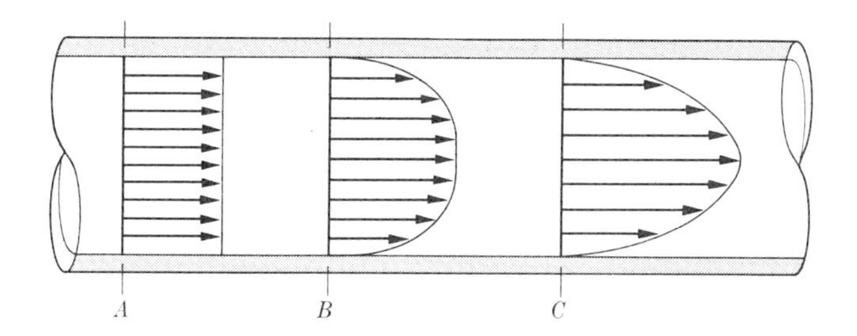

Fig. 14–6

In laminar pipe flow, as illustrated in Fig. 14–6, fluid flow approaching the entrance to the pipe has a flat velocity profile, A. As the fluid moves down the pipe, as at B in Fig. 14–6, the boundary layer effect begins to appear and the velocity profile starts to develop. At this point the fluid resembles that shown for turbulent flow in Fig. 14–3. By the time the fluid reaches C in the conductor the parabolic profile has developed. At this point the flow is said to be "fully developed." We can see that in order for the velocity distribution found in the boundary layer to exist, adjacent layers of fluid must be slipping past each other; that is, shearing of the fluid must occur. Since this is related to viscosity of the fluid, it can be seen that the viscous flow losses incurred in pipe flow occur in the boundary layer.

Whether or not flow in the boundary layer is laminar or turbulent depends upon the Reynolds number. It is interesting to note that in flow past a flat plate, as shown in Fig. 14–7, the initial flow is always laminar in the boundary layer. It may undergo a transition to turbulent flow at some point downstream from the leading edge of the plate. If this occurs, a discontinuity in the velocity profile is evident.

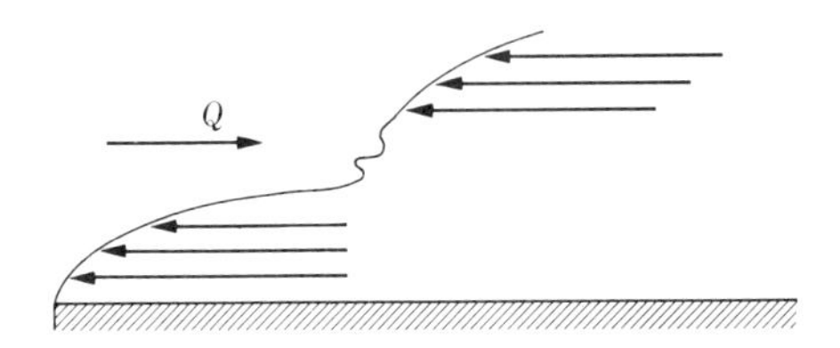

Fig. 14–7

The Reynolds number for a flat plate is given by the expression

$$N_{R_x} = \frac{v \cdot x}{\nu}, \tag{14–1}$$

where
v = flow velocity,
x = distance from leading edge of the plate, and
ν = kinematic viscosity.

Boundary-layer flow will be laminar for a Reynolds number up to about 500,000. Under laminar flow conditions the mean drag coefficient can be approximated by the relationship

$$C_d = 1.328/\sqrt{N_R}. \tag{14–2}$$

The boundary layer thickness at any distance from the leading edge can be determined from

$$\Delta/x = 5.20/\sqrt{N_{R_x}}, \tag{14–3}$$

where
Δ = thickness,
x = distance from the leading edge, and
N_{R_x} = Reynolds number at distance x from the leading edge.

If the flow past the plate is turbulent, the drag coefficient equation will have the form

$$C_d = 0.074/N_R^{0.20} \tag{14–4a}$$

when the Reynolds number lies between 2×10^5 and 10^7, or

$$C_d = 0.455/(\log_{10} N_R)^{2.58} \tag{14–4b}$$

when the Reynolds number lies between 10^6 and 10^9. Under turbulent flow conditions the boundary layer thickness is approximately

$$\Delta/x = 0.38/N_{R_x}{}^{0.20} \tag{14-5a}$$

when the Reynolds number lies between 5×10^4 and 10^6, or

$$\Delta/x = 0.22/N_{R_x}{}^{0.167} \tag{14-5b}$$

when the Reynolds number lies between 10^6 and 5×10^8.

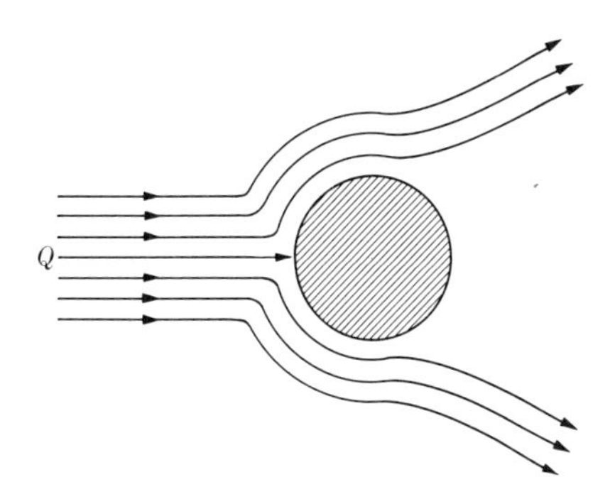

Fig. 14–8

The concept of boundary-layer effect can be extended to shapes other than flat surfaces. Thus the phenomenon has been an area of intensive study on the part of aerodynamicists because it is, in large measure, responsible for aerodynamic drag on aircraft and missiles. Figure 14–8 shows fluid flowing past a cylinder. Note that the streamlines must separate and pass around the cylinder. A boundary layer will develop along the surface of the cylinder as the fluid flows past it. At some point around the circumference of the cylinder the boundary layer lifts off the surface; this process is called *separation*. Many of the aerodynamic drag phenomena are related to this action.

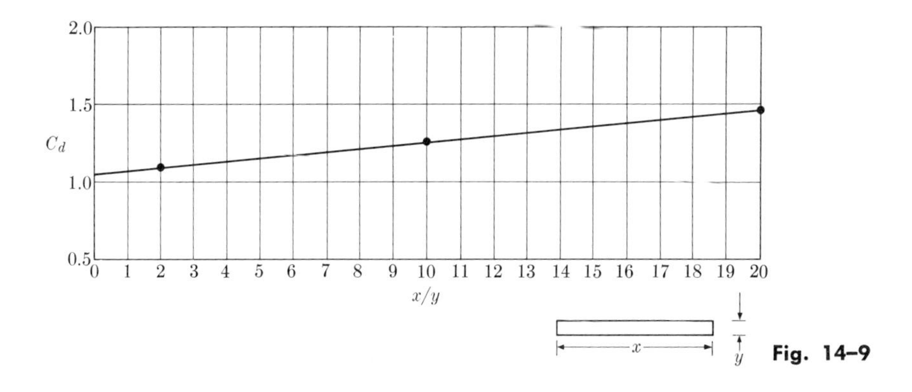

Fig. 14–9

In general, the drag exerted on a submerged body by fluid flowing past it is a function of the square of the velocity and the density of the fluid: $d = f(\rho, v^2)$. A constant of proportionality C_d changes this to the equation

$$\text{Drag} = d = C_d \rho A v^2/2, \tag{14-6}$$

where C_d is an empirically determined drag coefficient which varies for each shape. Typical drag coefficients for a flat plate are given in Fig. 14–9.

Flow Losses in Pipes

There are five general causes of loss of potential energy by a flowing stream as it passes through a conductor. One of them is considered the major loss; the other four are minor losses.

1) Major Loss
   a) Due to viscous-friction (shearing) effects associated with the flow of fluid through a pipe.
2) Minor Losses
   a) Due to the effects of a sudden contraction of cross section of the pipe.
   b) Due to a sudden expansion of cross section.
   c) Due to obstructions, such as pipe fittings, valves, etc.
   d) Due to curves and bends in pipe lines.

As a result of a great deal of experimental work on viscous losses in pipe flow, empirical expressions have been developed for calculating the magnitude of the losses which could be aniticipated in a given flow system. The general form of these equations is

$$h_f = f \frac{L}{D} \frac{v^2}{2g}, \tag{14-7}$$

where
h_f = head loss, in feet or inches,
L = length of conductor, in feet or inches,
D = diameter of pipe, in feet or inches,
v = velocity of flow,
f = a friction factor.

Note that this equation shows that the viscous losses in a pipe are a function of the velocity head. Thus flow losses are a function of the square of the velocity, and f is an empirically determined dimensionless friction factor. Many factors go into the determination of f; the major consideration is whether the flow is laminar or turbulent. Equation (14–7), using an empirically determined friction factor f, is for turbulent flow and is called the *Darcy-Weisbach* formula.

For flow velocities usually encountered in pipe flow, the friction factor f depends on two factors:

1) The Reynolds number.

2) The geometry or condition of the surface of the conductor.

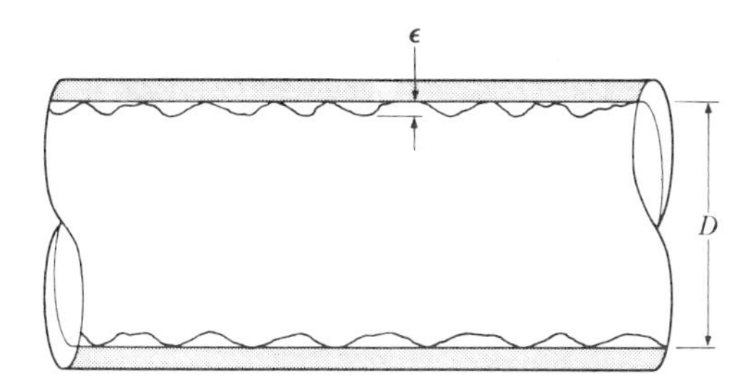

Fig. 14–10

The condition of the surface can be expressed by a parameter called the *relative roughness*, ϵ/D. Figure 14–10 illustrates the significance of this parameter, which may be thought of as the ratio between the height of the surface irregularities, ϵ, which is called the *absolute roughness*, and the diameter of the pipe.

It has been demonstrated experimentally that for laminar flows, i.e., Reynolds number less than 2000, $f = 64/N_R$. The expression for viscous losses then becomes

$$h_f = \frac{64}{N_R} \frac{L}{D} \frac{v^2}{2g}. \tag{14–8}$$

In this form, Eq. (14–8) is known as the *Hagan-Poiseville* formula, which indicates that for laminar flow, viscous losses are independent of surface roughness. Thus pipes of different surface conditions have the same losses for the same Reynolds number. The chart of Fig. 14–11 provides a means for estimating friction factors for flows at various Reynolds numbers as a function of relative roughness. The data presented in Fig. 14–12 describe absolute and relative roughness information for various types of commercial pipe.

A number of empirical formulas for calculating friction factors have been proposed:

1) For turbulent flow in smooth and rough pipes:

$f = 8\tau_0/\rho v^2, \quad \tau_0 =$ shear stress at pipe wall.

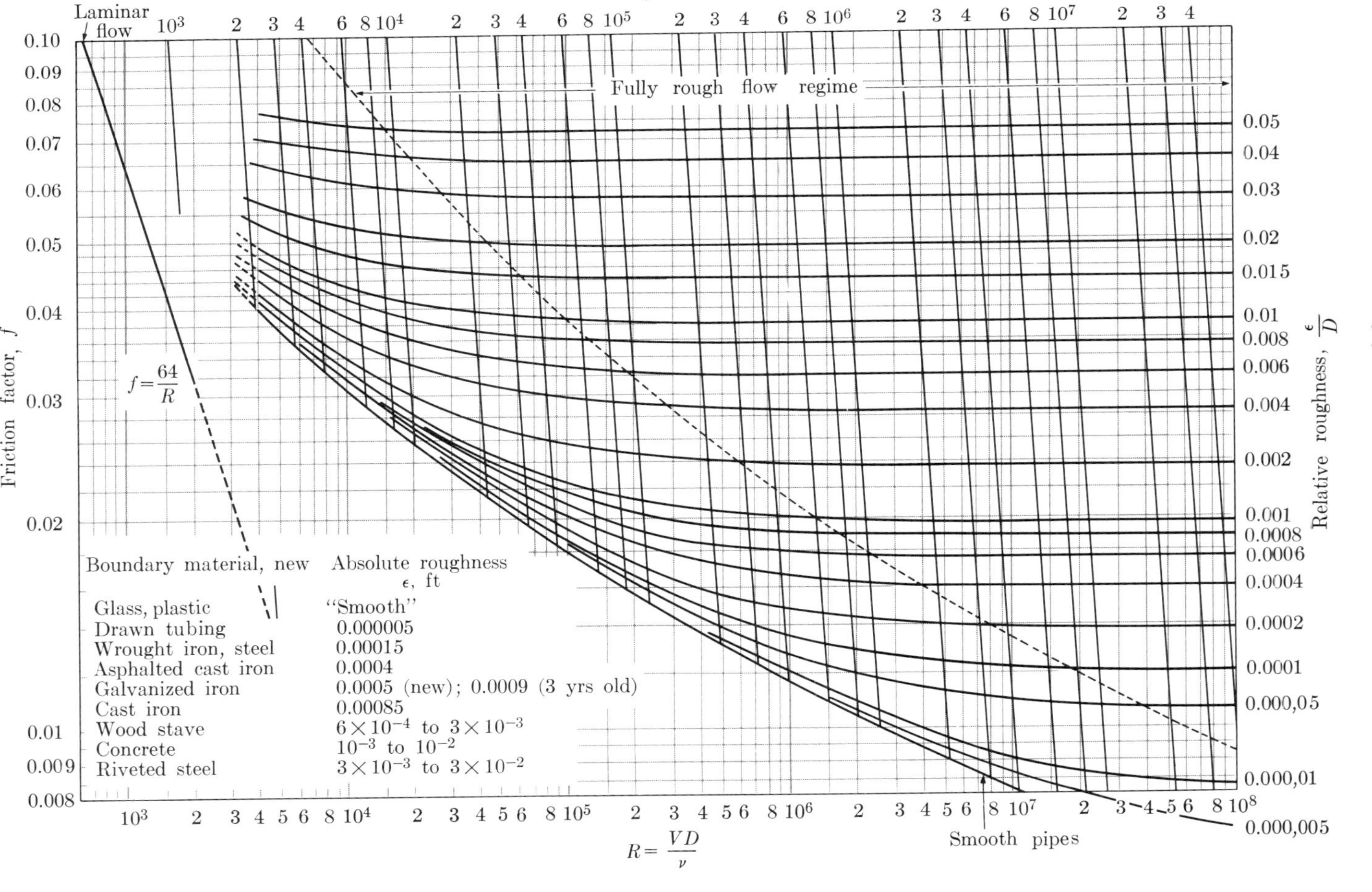

Fig. 14–11. [From L. F. Moody, "Friction Factors for Pipe Flow," *ASME Trans.* **66,** 8 (Nov. 1944).]

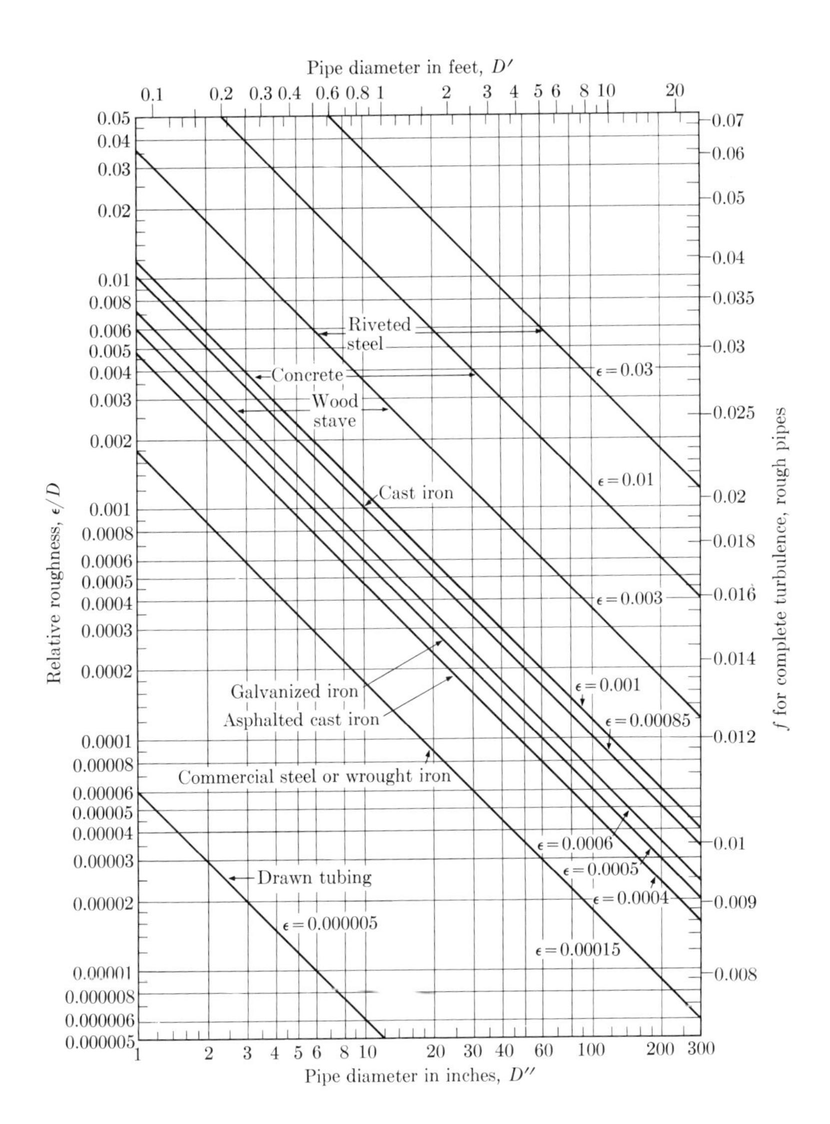

Fig. 14–12. [From L. F. Moody, "Friction Factors for Pipe Flow," *ASME Trans.* **66**, 8 (Nov. 1944).]

2) The Blasius equation for smooth pipes with Reynolds numbers between 3000 and 100,000: $f = 0.316/N_R^{0.25}$.

3) For smooth pipes at Reynolds numbers up to 3,000,000:

$$\frac{1}{\sqrt{f}} = 2 \log (N_R\sqrt{f}) - 0.8.$$

4) For rough pipes:

$$\frac{1}{\sqrt{f}} = 2 \log \frac{r_0}{\epsilon} + 1.74.$$

5) Colebrook equation for all pipes:

$$\frac{1}{\sqrt{f}} = -2 \log \left[\frac{\epsilon}{3.7d} + \frac{2.51}{N_R\sqrt{f}}\right].$$

EXAMPLE

A hydraulic oil with a specific gravity of 0.95 flows through a horizontal commercial steel pipe 2 in. in diameter. It is delivered by a pump at the rate of 3960 gpm. The temperature is 120°F. Calculate the pressure drop per 100 ft of pipe.

Solution.

Step 1. From Eq. (14–2) or (14–3) we can determine the parameters necessary to make this calculation. These are: Reynolds number, flow velocity (average), length of pipe, diameter of pipe.
All of the above must be expressed in consistent units.

Step 2. Reynolds number:

$$N_R = \frac{vD}{\nu},$$

$$v = \frac{Q}{A} = \frac{3960 \text{ gal/min} \times 231 \text{ in}^3\text{/gal}}{\pi D^2/4 \text{ in}^2 \times 60 \text{ sec/min}} = 4851 \text{ ips},$$

$$D = 2 \text{ in.}, \qquad \nu = 1.69 \times 10^{-2} \text{ (from fluid data) ft}^2\text{/sec;}$$

$$N_R = \frac{4851 \text{ in/sec} \times 2 \text{ in}}{1.69 \times 10^{-2} \text{ ft}^2\text{/sec}}.$$

This does not check out dimensionally.

$$N_R = \frac{[(4851 \text{ in/sec})/12 \text{ in/ft}] \times [2 \text{ in}/(12 \text{ in/ft})]}{1.69 \times 10^{-2} \text{ ft}^2\text{/sec}} = 3986.7$$

This expression does check out dimensionally.

Step 3. Since the Reynolds number is over 4000 (upper critical velocity), we know that we will have turbulent flow. We must, therefore, determine a friction factor for use in the Darcy-Weisbach equation.

a) From the data of Fig. 14–12, we know that the relative roughness is about 0.0009.

b) From the chart of Fig. 14–11, we can determine that for a relative roughness of 0.0009 and a Reynolds number of 3986.7, the friction factor f is 0.04.

Step 4. $$h_f = f\frac{L}{D}\frac{v^2}{2g} = 0.04 \times \frac{100\text{ ft}}{2/12} \times \frac{(4851/12)^2}{2 \times 32.2\text{ ft/sec}^2}$$

$$= 61{,}500\text{ ft of head/100 ft of pipe.}$$

Step 5. $p = 0.433\, S_g h_f = 0.433 \times 0.95 \times 61{,}500 = 2560$ psi drop.

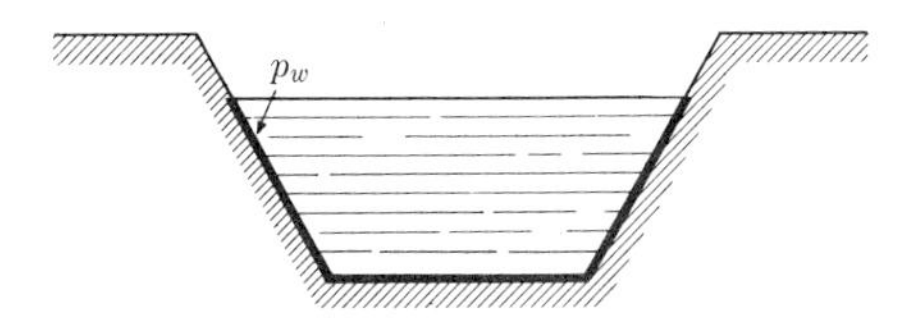

Fig. 14–13

Conductors of Noncircular Cross Section

The above discussions on pipe flow have assumed a uniform circular cross section for the flow conductor. In some instances, the cross section for the flow conductor is nonuniform and/or noncircular in section. The parameter used to define the configuration of the noncircular conductor is called the *hydraulic radius*. This parameter depends, in turn, on one called the *wetted perimeter*, which is the line of intersection of the wetted surface of a conductor and a cross-sectional plane, as shown in Fig. 14–13. The hydraulic radius is then defined as

$$R_h = \frac{\text{area of the cross section}}{\text{wetted perimeter}}.$$

In terms of the hydraulic radius, the Reynolds number is expressed by

$$N_R = \frac{4vR_h}{\nu}. \tag{14–9}$$

For a pipe flowing full, the wetted perimeter is πd, and the hydraulic radius is $d/4$.

IMPORTANT TERMS

Velocity profile is a plot of the variation in velocity across the section of a fluid conductor.

Average velocity is the mathematical average of all velocity vectors across the section of the conductor; it is the velocity implied in the continuity of flow equation.

Boundary layer is the layer of fluid in contact with the surface of a conductor such that the velocity in the layer is 99%, or less, of the maximum velocity in the flowing stream.

Drag is the force exerted on an immersed body by a fluid flowing past it, by virtue of the viscous shear effects of the fluid.

Flow losses are those losses of potential energy (head loss or pressure drop) resulting from the properties of flowing fluids and their interaction with the conducting system.

Darcy-Weisbach formula (or equation) is an empirically determined expression for calculating approximately the loss in potential energy.

Hagen-Poiseville equation is an empirical equation similar to the Darcy-Weisbach one, except that it is used to calculate losses for laminar flow conditions. (The Hagen-Poiseville equation can be analytically derived.)

Absolute roughness, ϵ, is the actual height of roughness projections on the surface of a pipe wall.

Relative roughness is the ratio of the absolute roughness to the diameter of the pipe. It is dimensionless.

Wetted perimeter is the length of the line of intersection of the cross section of the conductor and the wetted surface.

Hydraulic radius is the ratio of the cross-sectional area of the conductor and the wetted perimeter.

PROBLEMS

14–1 If laminar flow occurs between two flat parallel plates at an average flow velocity of 20 fps, what is the maximum velocity?

14–2 The maximum flow velocity in a 1-in. pipe in which laminar flow exists is 25 fps. Determine the flow rate of liquid in the pipe. If flow were turbulent, would you be able to answer this question?

14–3 Two-dimensional laminar flow exists between two fixed flat parallel plates spaced 8 in. apart. The maximum velocity is 30 fps. If the fluid is water, what is the pressure drop in the direction of flow for a length of 50 ft?

14–4 A flat plate 2 ft × 17 ft moves through water with a velocity of 35 fps. Calculate the drag on the plate.

14–5 The drag on a flat plate 4 ft × 32 ft moving through water in a direction parallel to its length is found to be 2000 lb. Calculate the velocity.

14–6 The plate of Problem 14–5 is moving through water at a velocity of 5 fps. Assuming that the boundary-layer flow is initially laminar, calculate the point at which the transition to turbulent flow takes place.

14–7 In the system of Problem 14–6, what is the thickness of the boundary layer at the point calculated?

14–8 Calculate the drag on the plate of Problem 14–6.

14–9 Calculate the critical velocity (lower) for hydraulic oil flowing through (a) a $\frac{1}{2}$-in. pipe; (b) a $\frac{3}{4}$-in. pipe; (c) a 1-in. pipe; (d) a 2-in. pipe; (e) $\frac{1}{2}$-in. tubing; (f) $\frac{3}{4}$-in. tubing; (g) 1-in. tubing.

14–10 To ensure laminar flow conditions, what size pipe must be used to transfer 50 gpm of phosphate ester fluid?

14–11 A rule of thumb of fluid power practice says to keep flow velocity to a maximum of 4–5 fps in the suction line of a pump, and about 20–25 fps in the discharge line. For water-in-oil emulsion fluid, calculate the pipe size required to ensure laminar flow in each line. If a pipe of the same size were used for both suction and discharge lines, would the flow remain laminar in the discharge pipe?

14–12 Calculate the friction factors for the conductors of Problem 14–9.

14–13 Hydraulic oil at 100°F flows through a 1-in. pipe at a rate of 40 gpm. Calculate the head loss per 100 ft of pipe.

14–14 A cast iron pipe 6 in. in diameter carries water (60°F) at a rate of 800 gpm. Calculate the head loss in 1000 ft of the pipe.

14–15 Water flows through a 400-ft length of $1\frac{1}{2}$-in. commercial steel pipe at a rate of 65 gpm. The downstream end of the pipe is 75 ft higher than the upstream end. Calculate the pressure difference between inlet and outlet.

14–16 A 12-in. steel pipe conducts water to a standpipe (water tower) in a municipal water system. The vertical height to the bottom of the storage tank is 120 ft. Determine the horsepower required to pump in water at a rate of 3000 gpm. If the pump is 70% efficient, what must be the drive-motor horsepower rating?

14–17 Oil lines at an unloading dock are steel pipes 8 in. in diameter. Fuel oil is loaded into tankers at a flow rate of 1200 gpm. The distance from the storage tanks to the side of the tanker is 3000 ft. The vertical lift to deck level is 50 ft. If the pumps are located at the storage tanks, what pumping horsepower would be required?

14–18 Ballast tanks on a submarine are located 50 ft apart at the same level in the hull. To trim the ship, sea water transfer pumps are located at each tank to transfer water from one to the other. If the submarine is listing 15° to port, what horsepower would be required to transfer 1000 gpm through a 6-in. steel tube to the starboard tank?

14–19 A fluid power pump delivers 100 gpm mineral-base hydraulic oil (120°F) through a 2-in. commercial pipe 75 ft long. What is the pressure drop through the pipe? What is the rate of heat generation?

14–20 Calculate the head loss for hydraulic oil (100°F) flowing through a $\frac{3}{4}$-in. commercial steel pipe: (a) at the lower critical velocity; (b) at the upper critical velocity. What is the percentage difference between the two? Discuss the significance of this difference relative to pressure drop and heat generation in a fluid-power system.

CHAPTER 15

Flow losses in pipes

Pipe Flow Losses

In Chapter 14, we described five sources of loss in potential energy in a fluid system. These losses show up as head losses or pressure drops, which are the two methods of expressing the same potential energy loss. The major loss is due to viscous shear effects as the fluid flows through the pipe and can be calculated by the Darcy-Weisbach or Hagen-Poiseville equation. A modification of the Reynolds-number formula allows us to use the parameters in the dimensional units in which they are most often encountered. This is

$$N_R = \frac{7740vD}{\nu}, \tag{15-1}$$

where v = velocity, ft/sec,
D = pipe diameter, in.,
ν = kinematic viscosity, centistokes.

A modification of the Darcy-Weisbach equation gives the loss directly in pressure drop, psi, rather than in feet of head loss. In fluid power systems the pressure drop is more meaningful than the head loss. This modification is

$$P = 0.0808f \frac{L}{D} v^2 S_g \text{ psi,} \tag{15-2}$$

where
f = friction factor,
L = length of conductor (ft),
D = diameter of conductor (in.),
v = average velocity of flow (fps),
S_g = specific gravity of fluid.

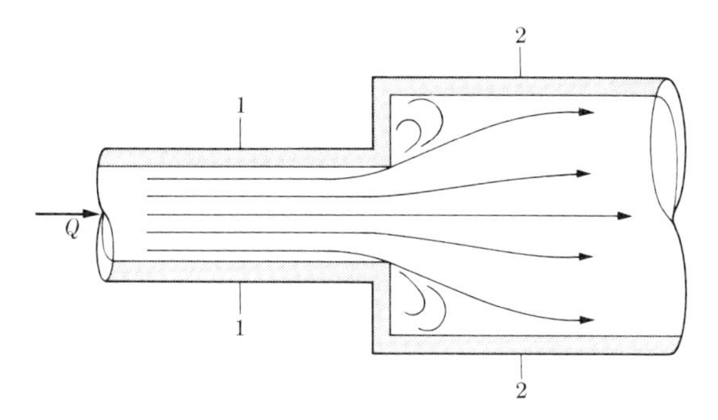

Fig. 15–1

Minor Losses

The loss due to the sudden enlargement of a pipe cross section is shown in Fig. 15–1. When there is such a sudden enlargement, the fluid, because of its momentum, cannot immediately make the sudden change in direction needed for the larger pipe diameter to be filled. Thus turbulence is developed in the "corners" of the enlarged section. It can be shown that in this case the head loss is a function of the velocity head:

$$h_f = \left(\frac{A_2}{A_1} - 1\right)^2 \cdot \frac{v_2^2}{2g} \tag{15-3}$$

The loss due to a sudden reduction of a cross section (Fig. 15–2) can be expressed as

$$h_f = K \frac{v_1^2}{2g}. \tag{15-4}$$

Remember that in Chapter 9 we indicated that a standard short tube behaved in a manner similar to the entrance to a pipe. The value of the constant K depends on the ratio of the pipe diameter at the large section to that at the small section. The K-values are empirically determined. We

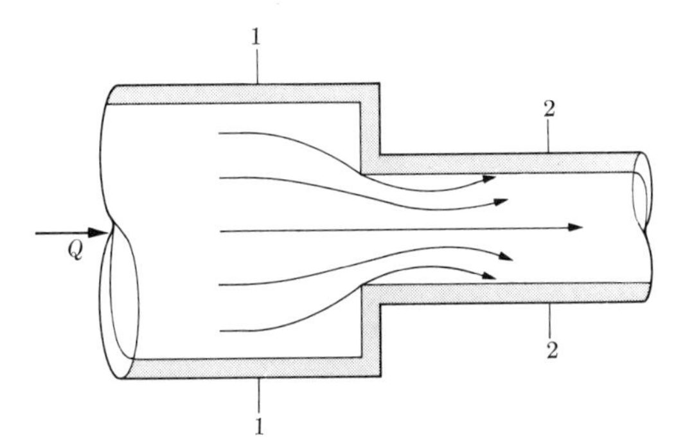

Fig. 15–2

list below several typical values of K for a number of ratios of d_1/d_2:

d_1/d_2	4	3.5	3.0	2.5	2.0	1.5	1.25	1.1
K	0.45	0.43	0.42	0.37	0.28	0.19	0.10	0

When d_1/d_2 is greater than 4.0, K equals 0.5.

Losses in pipe fittings, bends, etc., may also be expressed in the general form $h_f = K(v^2/2g)$. That is, the loss is a function of the velocity head. Values of K are empirically determined for the particular device under consideration. (The reader is also referred to the loss and flow calculations for various flow control devices given in Chapter 9.) Table 15–1 gives typical values of K for a number of such devices.

TABLE 15–1

Flow device	K	*Flow device*	K
Globe valve, wide open	10.0	Return bend	2.2
Angle valve, wide open	5.0	standard tee	1.8
Gate valve, wide open	0.19	standard elbow	0.9
Gate valve, $\frac{1}{4}$ closed	1.15	medium sweep elbow	0.75
Gate valve, $\frac{1}{2}$ closed	5.6	long sweep elbow	0.60
Gate valve, $\frac{3}{4}$ closed	24.0	45° elbow	0.42

Equivalent Length

The expression for head loss in a pipe due to fluid friction contains the length of the conductor as one of the parameters. It would be convenient if we could express all components in the conducting system in terms of length; the results obtained could then be used directly in the Darcy-Weisbach or Hagen-Poiseville equation. This is essentially what *equivalent length* is.

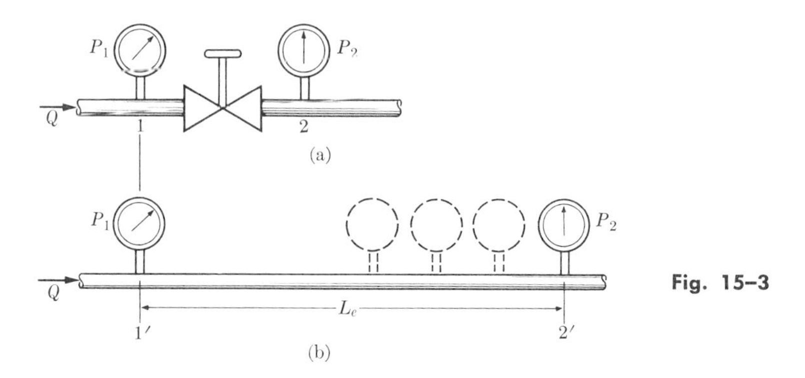

Fig. 15–3

Let us consider a flow device, such as a valve of a given size. See Fig. 15–3(a). A pressure gage is inserted at point 1 in the inlet to the valve, and a second one is placed at point 2 at the outlet. When a fluid flows through the valve, a pressure drop will be evidenced by the difference in readings between the two pressure gages. Now let us suppose that we have a length of pipe of the same nominal size as the valve. See Fig. 15–3(b). At point 1′, just opposite point 1 in Fig. 15–3(a), we insert a pressure gage. Suppose that we take a second pressure gage and insert it anywhere along the length of the pipe beyond point 2. We will find some point 2′ in the pipe where the reading of the second gage will be the same as that downstream from the valve. Thus the pressure drop through that length of pipe between points 1′ and 2′ will be the same as the pressure drop across the valve. The distance between points 1′ and 2′, L_e, is the equivalent length of the valve in Fig. 15–3(a). If the equivalent length is substituted in the Darcy-Weisbach formula, the head loss calculated would be the same as that which we would observe across the valve. As a practical matter, of course, we cannot go the route of trying to find experimentally the pipe length which will yield a pressure drop equal to that of a flow device. However, the following expression, based on empirically determined constants or K-factors, can be used to calculate an approximate equivalent length:

$$L_e = \frac{K}{f} D \qquad \text{if } D \text{ is in feet,} \tag{15–5a}$$

or

$$L_e = \frac{K}{f} \frac{D}{12} \qquad \text{if } D \text{ is in inches.} \tag{15–5b}$$

Note that the friction factor, which was previously discussed in reference to the Darcy-Weisbach equation, as well as the pipe diameter and the K-factor, is part of this expression.

EXAMPLE

Assuming that a hydraulic oil of specific gravity 0.95 flows through a 2-in. steel pipe at a rate of 66 gpm, determine the equivalent length for a globe valve placed in the line.

Solution.

Step 1. From Eq. (15–5), that is, $L_e = K/f \times D/12$, we know that to evaluate the equivalent length, we must find the K-factor, the friction factor, and the diameter.

Step 2. From Fig. 15–3, we see that the K-factor for a globe valve is 10 when the valve is wide open.

Step 3. Let us find the friction factor f:

a) $$N_R = \frac{vD}{\nu} = \frac{405 \times 2/12}{1.69 \times 10^{-2}} = 4000;$$

thus the flow is turbulent.

b) From Fig. 14–12, the relative roughness for the commercial steel pipe is $0.0009 = e/D$.

c) From Fig. 14–11, we determine that the friction factor is about 0.04 for a Reynolds number of 4000 and a relative roughness of 0.0009.

d) Thus

$$L_e = (10/0.04) \times \tfrac{2}{12} = 41.6 \text{ ft.}$$

Fig. 15–4

Flow In Branched-Pipe Systems

Consider a simple parallel pipe system, such as that shown in Fig. 15–4. If we place a pressure gage in the system at point 1 and another gage at point 2, the pressure difference, $P_1 - P_2$, will represent the pressure drop across the system. The drop will be a function of the flow rate Q. More important, note that the same pressure drop will be imposed across both branches of the parallel piping system; that is, the potential energy difference across branch 1 will be the same as the potential energy difference across branch 2. We have here a situation which is analogous to a parallel electrical circuit. If we impose a certain potential difference across such an electrical circuit, a current will flow through each branch as a function of the resistances of the branches. In the case of fluids we have, in effect, a potential difference existing across the fluid circuit. Because of the resistance of the conductors in the piping circuit to fluid flow, a certain flow rate, Q_1, will pass through branch 1; and a second flow rate, Q_2, will exist in the other branch. The total flow into the branched circuit must be equal to the total outflow; thus $Q = Q_1 + Q_2$. From our discussions of orifice flow and flow through orifice-like devices, we remember that the general form of the flow equation is $Q = AK_0\sqrt{\Delta P}$, where $K_0 = \sqrt{2g/\gamma}$. Since the flow rate is a function of the physical parameters of the system and the square root of the pressure difference, we have

$$Q_1 = A_1K_0\sqrt{\Delta P_1} \qquad \text{and} \qquad Q_2 = A_2K_0\sqrt{\Delta P_2}.$$

But $\Delta P_1 = \Delta P_2$. Solving the above expressions for ΔP and employing flow-continuity principles and Eq. (15–2), we obtain the following expression for pressure drop across the parallel network:

$$\phi_t = \left(\frac{A_v\sqrt{\phi_1\phi_2}}{A_1\sqrt{\phi_2} + A_2\sqrt{\phi_1}}\right)^2, \qquad \Delta P = \phi_t \cdot v^2,$$

where

$$\phi_1 = 0.0808\,\frac{L_1}{D_1}\,f_1S_g, \qquad \phi_2 = 0.0808\,\frac{L_2}{D_2}\,f_2S_g. \tag{15–6}$$

To solve for the individual branch flows, it therefore becomes necessary to know the parameters of the two branches, calculate the pressure drop, and resort to one of the expressions utilizing pressure drop to compute flow rate. When this is attempted an immediate problem will become evident. The Reynolds number must be calculated, but we do not know the flow velocity. Since the Reynolds number is a dimensionless ratio expressing the relationship of a number of factors related to fluid flow, there ought to be other such ratios which could be found by dimensional analysis. There are two such useful ratios: one does not involve the velocity, and the other does not use the diameter. The first is

$$S = \frac{((\Delta p/L)\cdot D^3\rho)^{1/2}}{\mu} = N_R\left(\frac{f}{2}\right)^{1/2} \tag{15–7}$$

The second is

$$T = \frac{(Q^3(\Delta p/L)\rho^4)^{1/5}}{\mu} = \frac{N_R}{4}\,(8\pi^3 f)^{1/5}. \tag{15–8}$$

Note that the quantity of p/L is equal to the pressure gradient or the change in pressure per unit length of pipe. To utilize these expressions, we must plot sets of curves of constant S or T as a function of N_R and f. Then S or T for the pipe system at hand can be calculated, and a corresponding N_R can be determined from the above curves.

In general, the procedure for solving branched-pipe flow systems requires the writing of the necessary number of simultaneous equations or employing empirically determined modifications of the Darcy-Weisbach equation in which the coefficient depends upon only relative roughness of the pipe. Several such formulas have been devised by Manning, Schoder, Scobey, Hazen-Williams, etc., and will be found in the more detailed literature on pipe-flow systems.

IMPORTANT TERMS

Major loss in a pipe-flow system is that due to viscous-friction effects because of the flow of the fluid through the pipe.

Minor losses in a pipe system are those due to sudden enlargement or reduction in cross section, and to bends, elbows, and fittings, etc.

K-value is the experimentally determined constant of proportionality between the head loss in a fluid system and the velocity head.

Equivalent length of a fitting, valve, bend, or other device contributing to flow losses is the length of a piece of pipe of the same nominal size which, under the same flow conditions, would have the same head loss or pressure loss as that observed in the flow-resisting device.

Branched pipe is a piping system consisting of one pipe split into two or more parallel branches through which the fluid can flow.

PROBLEMS

15–1 What is the head loss through a 2-in. 90° ell when water flows through it at a rate of 200 gpm?

15–2 What is the head loss through a 1-in. angle valve at a flow rate of 60 gpm?

15–3 Calculate the head loss through a standard $1\frac{1}{2}$-in. tee at a flow rate of 120 gpm.

15–4 A conductor of diameter $D_1 = 2$ in. is connected to a reservoir wall as shown in Fig. 15–5. The conductor expands to a diameter of $D_2 = 4$ in. $L_1 = 150$ ft, $L_2 = 100$ ft. Water flows out of the reservoir at a rate of 250 gpm. (a) Calculate the entrance loss to the 2-in. section of pipe. (b) Calculate the loss due to sudden expansion of the conductor. (c) Calculate the pressure drop through the system.

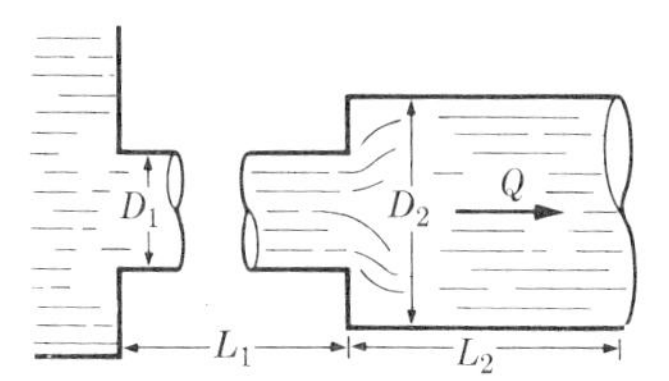

Fig. 15–5

15–5 If $D_1 = 1$ in., $D_2 = 3$ in., and $Q = 60$ gpm in Fig. 15–5, what is the loss due to sudden enlargement? What is the entrance loss?

15–6 What would be the entrance loss and loss due to sudden *contraction* of the conductor in Fig. 15–5 if $D_1 = 4$ in., $D_2 = 2$ in., and $Q = 200$ gpm?

15–7 A globe valve is placed in a 100-ft run of 1-in. commercial steel pipe of standard weight. Water flows through at a rate of 55 gpm. What is the equivalent length of the pipe and valve? What is the head loss?

15–8 A pipe run consists of 20 ft of $\frac{3}{4}$-in. commercial steel pipe, a 45° elbow, a 40-ft run of $\frac{3}{4}$-in. pipe, another 45° elbow, a half-open gate valve, and 10 ft of $\frac{3}{4}$-in. pipe. What are the equivalent length and the pressure drop in the run if hydraulic oil flows through it at a rate of 35 gpm?

15–9 In Problem 15–8, how much horsepower would be required to compensate for the loss? How many Btu (heat) would be generated?

15–10 A 6-in. steel pipe line in an oil tank farm is 500 ft long. It contains three 90° elbows, two tees, two 45° elbows and two half-closed gate valves. The flow rate is 1000 gpm. Determine the pumping horsepower required to move the oil through the line. How many Btu/min are generated? Where does this heat go?

15–11 If the pump of Problem 15–10 is 80% efficient, what size drive motor is required?

15–12 A compound pipe system is made up of 1000 ft of 2-in. steel pipe and 3000 ft of 4-in. steel pipe in series. The fluid is water. Its flow rate is 250 gpm. Calculate the pressure drop across the 4000 ft of the system.

15–13 A compound piping system is made up of 1500 ft of 2-in. cast iron pipe with a standard tee and contains a wide-open globe valve. This system is connected in series with 2000 ft of 3-in. cast iron pipe (two 45° elbows) and 4000 ft of 4-in. pipe with a standard tee and another globe valve. $Q = 100$ gpm of water. Calculate the head loss across the entire system. What would be the pumping horsepower required to overcome these losses? How many Btu/min (heat units) would be generated?

15–14 Consider a branched-pipe system like that shown in Fig. 15–4. If each branch in the parallel circuit consists of 20 ft of $\frac{1}{2}$-in. steel tube, what is the pressure drop across the system for hydraulic oil flowing at a rate of 18 gpm? Inlet and outlet conductors are made of $\frac{3}{4}$-in. steel tubing.

15–15 The Hazen-Williams formula is frequently used to solve pipe flow problems. It is expressed as

$$v = 1.318 C_1 R^{0.63} S^{0.54},$$

where

v = flow velocity, fps,

R = hydraulic radius,

S = slope of the hydraulic gradient (piezometric line),

C_1 = Hazen-Williams coefficient of relative roughness.

It can be seen that the solution is not a simple one; this is why most of the parameters are tabulated in handbooks for simplified calculations. Assume that we have 1000 ft of 8-in. pipe. From the tables given in handbooks we find that $C_1 = 100$. A head loss of 48 ft is observed. Calculate the flow rate through the pipe.

CHAPTER 16

An example illustrating the calculation of pipe losses

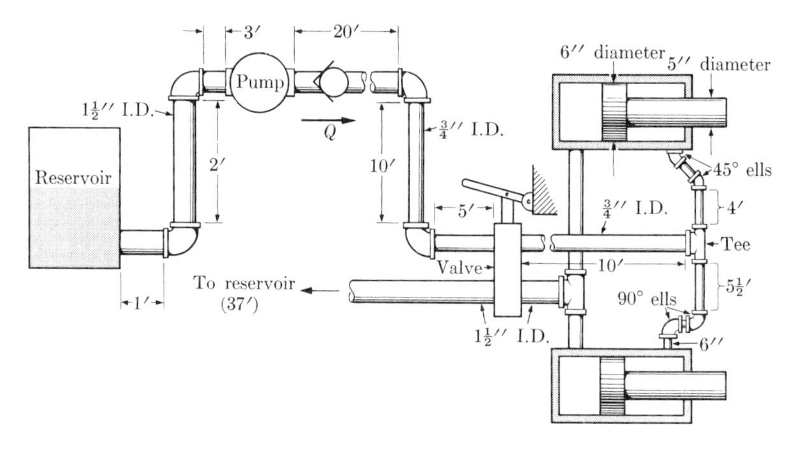

Fig. 16–1

Sample Fluid Power Circuit

Figure 16–1 illustrates a typical fluid power circuit for which calculations of pressure loss will be made. The positive-displacement gear pump is frequently used in fluid power applications. The pump is driven by a 15-hp electric motor and has an overall efficiency of 75%. The discharge pressure of the pump is 1000 psi. Oil of a viscosity of 110 centistokes at 120°F ($S_g = 0.88$) is being used in the system. We proceed as follows:

Step 1. Since the prime mover is a 15-hp electric motor and the pump is 75% efficient, the hydraulic horsepower delivered to the system equals $15 \times 0.75 = 11.25$ hp.

Step 2. Next we determine the flow rate. From Eq. 4–2 for hydraulic horsepower, we obtain

$$\text{hp} = PQ/1714,$$

$$Q = 1714\ \text{hp}/P = 1714 \times (11.25/1000) = 19.3\ \text{gpm}.$$

Step 3. From the principle of flow continuity, $Q = Av$ or $v = Q/A$, we have

$$v = \frac{(\text{gpm})\ (\text{in}^3/\text{gal})}{(\text{in/ft})\ (\text{sec/min})\ (\text{in}^2)} = \frac{19.3 \times 231}{12 \times 60 \times 1.77} = 3.32\ \text{ft/sec}.$$

Step 4. Next we determine Reynolds number:

$$N_R = 7740\ vD/\nu = 7740 \times 3.32 \times (1.5/110) = 350.$$

Thus the flow will be laminar.

Step 5. We see that the flow will be laminar since Reynolds number falls below the critical value of 2000. With laminar flow, the friction factor $f = 64/N_R = 64/350 = 0.183.$

We should mention at this point that it has been common practice in fluid power technology to try to keep flow velocities in the inlet lines to pumps to a maximum of 4 or 5 ft/sec, because it is desirable to maintain laminar flow in the suction or inlet to the pump.

Step 6. Our next step is the calculation of losses. We begin with the losses on the suction line, and we determine first the equivalent lengths of the various parts of the line.

a) Suction line:

1) Pipe entrance loss:

$$L_e = \frac{K}{f}\frac{D}{12} = \frac{0.5}{0.183}\frac{1.5}{12} = 0.342\ \text{ft}.$$

2) 90° elbow:

$$L_e = \frac{K}{f}\frac{D}{12} = \frac{0.9}{0.183}\frac{1.5}{12} = 0.615\ \text{ft} \times 2 = 1.33\ \text{ft}.$$

3) Straight pipe. The equivalent length of pipe is, of course, the actual length: $1 + 2 + 3 = 6$ ft.

4) The total equivalent length of the suction line is the sum of (1) through (3):

$$L_e = 0.342 + 1.33 + 6 = 7.672\ \text{ft}.$$

5) Pressure drop:

$$P_f = 0.0808f\,\frac{L}{D}\,v^2 S_g = \frac{0.0808 \times 7.672 \times (3.32)^2 \times 0.88 \times 0.183}{1.5}$$

$$= 0.735 \text{ psi};$$

$$h_f = 2.32\,P_f/S_g = 2.32 \times 0.735/0.88 = 1.94 \text{ ft of head.}$$

6) It may be interesting to write Bernoulli's equation from the reservoir to the inlet port of the pump to determine the actual pressure at the inlet:

$$\frac{P_1}{\gamma} + \frac{v_1^2}{2g} + H_1 = \frac{P_2}{\gamma} + \frac{v_2^2}{2g} + H_2 + H_L,$$

$$0 + 0 + 0 = P_2/\gamma + 11 + 2 \text{ ft} + 1.94 \text{ ft},$$

$$P_2/\gamma = -14.94 \text{ ft}, \qquad P_2 = -5.17 \text{ psi}.$$

This brings out another aspect of fluid power practice. The distance through which the fluid is raised, as well as the flow velocity in the pipe, is limited. If the suction head (distance through which we try to raise the fluid plus the velocity head plus the head loss in the pipe) is high enough to reduce the suction pressure below the vapor pressure of the fluid or to a level where dissolved gases come out of solution, then a phenomenon known as *cavitation* can result. This is very damaging to pumps.

b) The calculations of losses for the other sections of the line in the circuit are made in a manner similar to that shown for the suction line. The reader should proceed with these calculations and complete the problem on his own.

c) Losses in direction control valves. Fluid power control valving, such as the direction-control (or circuit-switching) valve shown in Fig. 16–1, consists of devices having complex flow paths. In some simpler cases, the relationships of Chapter 9 may be used to approximate the losses through valves, etc. For more complex components, the manufacturer must furnish information on ΔP versus Q. For pneumatic valves, as contrasted with hydraulic valves, manufacturers frequently specify a C_v factor such that $Q = f(C_v, \sqrt{\Delta P})$. In either case, the pressure drop across the valve is determined as a function of flow rate through it and contributes to the total losses (or drops) in the system.

Another important consideration is the action of the ram (cylinder with large-diameter rod) used in the circuit. As shown in Fig. 16–2, a differential area exists across the piston from the blank end (no piston rod) to the rod end. Thus when oil is ported to the blank end at the rate of 19.3 gpm, as previously calculated, the return flow of oil from the rod end is relatively small; that is, the return flow equals the area ratio times the incoming flow rate, or $(8.7/28.3) \times 19.3 = 5.92$ gpm. If the pipe lines running to both the blank and rod ends of the cylinder are of the same size, then the flow velocity in the return line is only about one-third that in the inlet line. The flow losses would be about one-ninth as great.

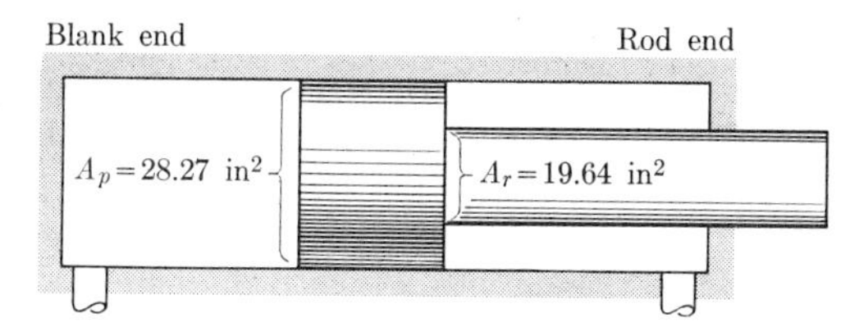

Fig. 16–2

On the other hand, when the direction control valve is switched so as to reverse the direction of flow, the 28.3 gpm is then ported to the rod end. Thus the fluid enters the smaller volume at the same rate at which it entered the larger volume on the blank end. The rate at which the return fluid is displaced from the blank end under these conditions is again a function of the area ratio, in this case, however, it is the inverse area ratio. The return flow rate will now be $(28.3/8.7) \times 19.3 = 63$ gpm. The flow velocity in the return line will be three times the velocity of the flow from the pump, and the losses will be about nine times as great.

Branch Lines

a) The lines leading from the direction control valve to the rams are $\frac{3}{4}$-in. pipes. They branch into two parallel lines where the tee is inserted to route oil to each of the rams.

b) Branching equations have been developed which show that the velocity in one branch can be calculated on the basis of the parameters of the system. Thus

$$v_1 = \frac{v \cdot L_2 \cdot A}{(A_2^2 L_1/A_1) + L_2 A_1}, \qquad v_2 = \frac{v \cdot L_1 \cdot A}{(A_1^2 L_2/A_2) + L_1 A_2}, \tag{16–1}$$

where quantities with subscript 1 refer to branch 1; those with subscript 2, to branch 2; and those with no subscript refer to the inlet or feeder line to the branches.

c) If the pipe sizes are the same, as is usually the case in fluid power circuits, then $A = A_1 = A_2$. In this case, the expressions in (b) reduce to

$$v_1 = \frac{vL_2}{L_1 + L_2}, \qquad v_2 = \frac{vL_1}{L_1 + L_2}. \tag{16–2}$$

d) We must calculate the flow velocity in the pipe as follows:

$$v = \frac{19.3\ (\text{gpm}) \times 231\ (\text{in}^3/\text{gal})}{12\ (\text{in/ft}) \times 60\ (\text{sec/min}) \times 0.44\ \text{in}^2} = 14.1\ \text{fps}.$$

e) Once we have determined the flow velocity we can calculate Reynolds number:

$$N_R = 7740 \times 14.1 \times (0.75/110) = 742;\ \text{laminar flow exists.}$$

f) Next we can calculate the friction factor:

$$f = 64/N_R = 64/742 = 0.086.$$

g) In order to proceed with the analysis of the losses in the circuit, we must determine the equivalent lengths of each of the branches. Before we can do this, flow velocities and Reynolds number must be determined for each branch.

1) We can use Eq. (16–2) to calculate the flow velocity:

$$v_1 = \frac{6 \times 14.1}{10} = 8.46\ \text{fps}, \qquad v_2 = \frac{4 \times 14.1}{10} = 5.64\ \text{fps}.$$

2) Next we calculate Reynolds number for each branch:

$$N_{R_1} = \frac{7740 \times 8.46 \times 0.75}{110} = 446;\ \text{laminar flow exists;}$$

$$N_{R_2} = \frac{7740 \times 5.64 \times 0.75}{110} = 298;\ \text{laminar flow exists.}$$

3) We can calculate the equivalent length of Branch 1.

Step 1. Calculate the friction factor:

$$f = 64/N_R = 64/446 = 0.143.$$

Step 2. Calculate equivalent length of the 45° elbows:

$$K = 0.42; \quad L_e = (K/f) \cdot (D/12) = \frac{0.42 \times 0.75}{0.143 \times 12}$$
$$= 0.1835 \times 2 = 0.367\ \text{ft}.$$

Step 3. An equivalent length for the exit loss is calculated next:

$$K = 1; \quad L_e = \frac{1 \times 0.75}{0.143 \times 12} = 0.437\ \text{ft}.$$

Step 4. The equivalent length of the straight pipe is the same as the actual length of pipe:

$$L = 4 \text{ ft.}$$

Step 5. The total equivalent length is the sum of the equivalent lengths calculated above:

$$4 + 0.367 + 0.437 = 4.804 \text{ ft.}$$

4) We can calculate the equivalent length of Branch 2.

Step 1. Calculate the friction factor:

$$f = 64/298 = 0.214$$

Step 2. Calculate equivalent length of the 90° elbows:

$$K = 1.8; \quad L_e = \frac{1.8 \times 0.75}{0.214 \times 12} = 0.526 \text{ ft.}$$

Step 3. An equivalent length for the exit loss is calculated next:

$$K = 1; \quad L_e = \frac{1 \times 0.75}{0.214 \times 12} = 0.292 \text{ ft.}$$

Step 4. The equivalent length of the straight pipe is the same as the actual length of pipe:

$$L = 6 \text{ ft.}$$

Step 5. The total equivalent length is the sum of the equivalent lengths calculated above:

$$L_e = 6 + 0.526 + 0.292 = 6.818.$$

5) It is important to note that the equivalent lengths we have just calculated will have an affect on the actual flow velocities. We must recalculate flow velocities using Eq. (16–2) but substituting the equivalent lengths of each branch for the actual lengths used previously.

(a) The flow velocity in branch 1 can be calculated using Eq. (16–2) and the equivalent length of branch 1:

$$v_1 = \frac{6.818 \times 14.1}{11.622} = 8.25 \text{ fps.}$$

(b) The flow velocity for branch 2 can be calculated using Eq. (16–2) and the equivalent length of branch 2:

$$v_2 = \frac{4.802 \times 14.1}{11.622} = 5.82 \text{ fps.}$$

(c) Next we must determine Reynolds number for the branches using the revised flow velocities calculated above:

$$N_{R_1} = \frac{7740 \times 8.25 \times 0.75}{110} = 835; \text{ laminar flow exists;}$$

$$N_{R_2} = \frac{7740 \times 5.82 \times 0.75}{110} = 307; \text{ laminar flow exists.}$$

(d) We can calculate the friction factor for each of the branches of the circuit:

$$f_1 = 64/835 = 0.0767,$$
$$f_2 = 64/307 = 0.209.$$

6) We can calculate the pressure drop in the branched circuit using the branching equations.

(a) The first step is to determine the value of ϕ_1 for branch 1 and ϕ_2 for branch 2:

$$\phi_1 = 0.0808 \frac{L_1}{D_1} f_1 S_g = 0.0808 \times \frac{4.802}{0.75} \times 0.0767 \times 0.88$$
$$= 0.0349,$$

$$\phi_2 = 0.0808 \frac{L_2}{D_2} \times f_2 \times S_g = 0.0808 \times \frac{6.818}{0.75} \times 0.209 \times 0.88$$
$$= 0.135.$$

(b) Using the values for ϕ_1 and ϕ_2 we can calculate ϕ_t for the branched circuit:

$$\phi_t = \left[\frac{\phi_1 \phi_2}{\phi_1 + \phi_2}\right]^2 = \left[\frac{0.0349 \times 0.135}{0.0349 + 0.135}\right]^2 = 0.0168.$$

(c) Then $p_f = \phi_t \times v^2 = 0.0168 \times 14.1 = 0.237$ psi.

The techniques discussed in this chapter are applicable to pipe flow systems in general. As has been previously indicated in this book, much of the application of the principles of fluid mechanics to actual practice is empirical; that is, it is based on experimentally verified equations and constants evaluated by tests. One of the reasons for this is the large number of variables encountered in even the most elementary systems, which places an exhaustive dissertation on pipe flow beyond the scope of this text. The reader who desires more detailed information is referred to the literature in the field of fluid mechanics.

IMPORTANT TERMS

Positive displacement pump is a pump design in which the fluid is allowed to enter the pump through an inlet port, is picked up by the pumping element, and carried through a cycle during the course of which it is momentarily isolated from the outside environment, and is finally forcibly ejected from the pump into the discharge pipe. Energy transfer occurs via potential energy change, rather than kinetic energy change.

Direction control valves are fluid power components whose functions are to direct, or switch, the fluid to the pipe lines through which it is to pass.

Cylinder or ram is a type of fluid power motor in which the mechanical output is a linear thrust, whose length—usually the piston-rod–cylinder-barrel length—is limited by the design of the unit.

PROBLEMS

16–1 Figure 16–3 is the fluid power circuit schematic of Fig. 16–1 of the text. The drawing uses the graphic symbols for fluid power systems developed by the American Standard Association. Using the same flow parameters as in the example problem in the text, carry out a complete analysis of the system illustrated in the above diagram.

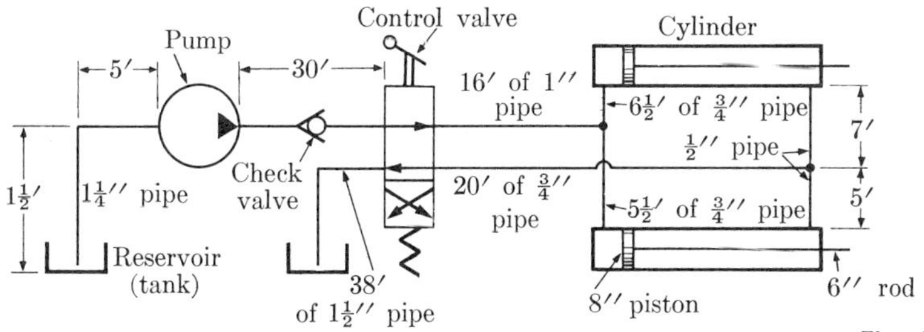

Fig. 16–3

16–2 Figure 16–4 illustrates a circuit diagram for a simple fluid power system. A 25-hp electric motor drives a fluid power pump which is 80% efficient. The fluid is a typical mineral-base hydraulic oil. Analyze the circuit from

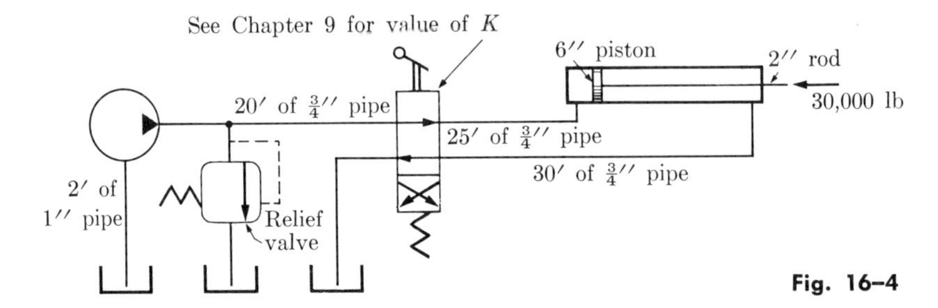

Fig. 16–4

the point of view of pressure loss in the system. Determine the heat generation rate. Considering the results of the analysis, do you believe that it is possible for the system to perform against the 30,000-lb load reaction indicated, if the piston velocity is 0.36 fps? If not, what load *could* it sustain?

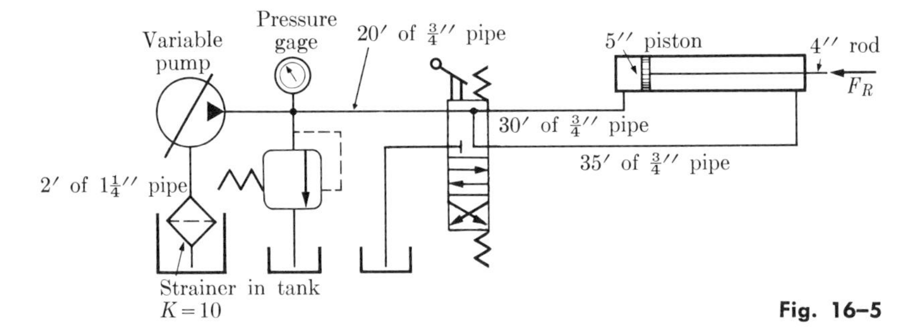

Fig. 16–5

16–3 Figure 16–5 shows a circuit diagram involving a regenerative cylinder (see Chapter 2). Assume that a 25-hp electric motor drives an 85% efficient fluid power pump. Mineral-base fluid is used. The pump discharge pressure is 1200 psi. Analyze system losses, determine the rate of heat generation, and calculate the load reaction, F_R, which could be handled by this system. At what constant velocity could the load be moved by the cylinder?

16–4 Repeat the analysis of Problem 16–2, assuming that the control valve is shifted and flow goes to the end of the cylinder which is opposite to that selected in Problem 16–2. What percent difference in losses do you observe? What part of the system contributes most to the change in losses with direction of flow through the valve?

16–5 Repeat Problem 16–3, assuming that the control valve is shifted *out* of the regenerative flow position and *into* the simple flow condition of pressure oil to the rod end of the cylinder, with cap end oil returned to tank through the control valve.

CHAPTER 17

Topics in compressible fluid power systems

Introduction to Compressible Flow

The bulk of the material presented so far has been concerned with the mechanics of incompressible fluids. The emphasis has been on industrial applications of fluid mechanics and, in particular, on the newly emerging fluid power technology.

The majority of introductory texts on fluid mechanics deal with noncompressible flow, because the new concepts which must be mastered to understand fluid mechanics are more readily understood when they relate to fluids which do not change. Thus we can consider the fluid to be a constant medium and need only treat the effects of changes of conditions, such as pressure, velocity, acceleration, etc. We can develop the laws which govern the reaction of the fluid and the reaction on other bodies from the simplified frame of reference of a nonchanging fluid medium.

The principles outlined so far in this text *are* those which are used in most industrial engineering situations. We would be remiss, however, if we did not at least mention the fact that fluid mechanics is much broader in scope and much more sophisticated in concept than is indicated by the elementary presentations contained in this volume. Theoretical fluid mechanics as developed in recent years under the impetus of the work of aerodynamicists, jet propulsion engineers, aerospace engineers in such fields as plasma jets and magnetohydrodynamics is among the most complex and sophisticated areas of our modern technology. Such work is far beyond the scope of an introductory text, but we should be aware of its existence.

Another new area of activity which is gaining momentum in industry is fluid power technology.

> **Fluid power** technology deals with the transfer, control, and storage of energy by means of a fluid in a closed system.

Note that the definition says *fluid*—not just liquid. Thus in dealing with fluid power systems we must be in a position to handle both liquid and gaseous systems, since both are used in the process of transmitting energy.

We have concentrated on liquid fluid mechanics throughout most of this text, for it is liquids which are considered to be noncompressible. Under most conditions this premise is close enough to the truth to prove satisfactory. In some fluid power systems we must recognize that liquids, too, are compressible to some small degree. It is fortunate that the mechanics of compressible fluids—gases—can be considered from a noncompressible standpoint for much of the analytical work relating to fluid-power systems. This greatly simplifies the problems involved.

Gas Laws

In Chapter 2, we developed the concept of *absolute zero pressure*, i.e., the concept of no atmospheric pressure whatsoever, or the complete absence of any pressure. We did this without going any deeper into what conditions might have to be satisfied to arrive at such a state. In reality, the concept of absolute zero pressure is only meaningful when considered in relation to the temperature of the fluid. In a gas, in particular, the state of absolute zero pressure presupposes a state of absolute zero temperature. Absolute zero temperature is equivalent to −273°C or −460°F. At that temperature, there is theoretically no molecular energy. Since gas pressure is a function of molecular energy, pressure ceases to exist when energy goes to zero.

To convert from the usual temperature scales to absolute temperature, we use the following formula:

$$T \text{ (absolute)} = T_f + 460, \quad \text{where } T_f = {}^\circ\text{F}.$$

When dealing with the gas laws, we must use not only absolute temperature but also absolute pressure. To convert from gage pressure to absolute pressure, we use the formula

$$P_a = P_g + 14.7,$$

where P_g = gage pressure in psi.

The following discussion summarizes the basic gas laws which describe the behavior of compressible fluids under changing conditions of pressure and/or temperature.

Charles' Law

1) If we hold the volume constant (by containing the gas), we have

$$P_1/P_2 = T_1/T_2. \tag{17–1}$$

This law tells us that with constant volume the absolute pressure is directly proportional to the absolute temperature.

2) If we hold the pressure constant and allow the volume to change, we have

$$V_1/V_2 = T_1/T_2. \tag{17–2}$$

Boyle's Law

If we hold the temperature constant ($T = °K$), we have

$$V_1/V_2 = P_2/P_1 \qquad \text{or} \qquad P_1V_1 = P_2V_2; \tag{17–3}$$

that is, the volume is inversely proportional to the pressure.

General Gas Law

The mathematical statement of the general gas law is

$$pV = MRT,$$

where p = absolute pressure, psf,
V = volume, ft^3,
M = mass of air,
T = absolute temperature,
R = universal gas constant = 53.3 lb-ft/$lb_m \times °F$ (for air).

The universal gas constant represents the amount of work (ft-lb) required to raise the temperature of one pound mass of air 1°F

$$R = 53.3/778.2 = 0.0686 \text{ Btu/lb}_m \cdot °\text{F}.$$

[*Note:* The volume of 1 lb of air at standard conditions is 12.39 ft^3/lb = V_a.]

Compressors

The fluid power engineer in industry encounters an interesting division of assignments, or interests. In hydraulic systems (those involving liquids), the energy transfer mechanism (the pump) is a prime part of the assignment for design or application. This is due to several reasons.

(1) Since line losses are considered to be excessive for central hydraulic systems, each machine has its own pump as an integral part of its hydraulic system.

(2) Control of the pump often is part of the over-all machine control.

This is not the case with pneumatic, or gas, fluid power systems. The usual practice is to have a central compressor plant, somewhere in the vicinity of the factory. Pressurized air is piped throughout the entire plant, and the fluid power engineer merely makes a connection to one of these distribution lines in order to get his source of energy. The approach is much the same as that for electricity. The engineer utilizing electric power for machine drive and control is generally not concerned with its generation.

Thus the fluid power engineer is usually not an expert in the compressor field. He looks upon these devices as "infinite" energy sources and just utilizes the compressed air wherever he wishes in the plant.

Pneumatic Fluid Power Systems

It is assumed that the central compressor facility will be capable of supplying a continuous flow of air at some minimum pressure which is always above that to be used at the machine. Figure 17–1 illustrates a typical pneumatic system connected into the air lines. A pressure *regulator* is almost invariably interposed between the source (the air main) and the machine on which the air is to be used. A regulator is a throttling device which maintains a relatively constant output pressure at some level lower than the input pressure. Within the limitations of the device, we may consider that the supply line to the machine itself is under constant pressure. From Boyle's law, we know that in this case the specific volume should remain constant also.

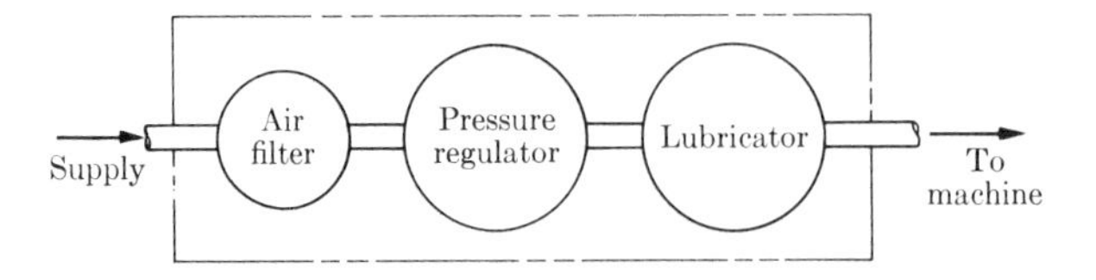

Fig. 17–1

Even if we use the pressurized air to move a cylinder, the unit volume will be constant so long as the pressure remains essentially constant. Under these conditions we can treat the flow of the gas as noncompressible flow.

EXAMPLE

A 10-in. bore cylinder is required to move a load resistance of 7060 lb through a distance of 30 in. in 5 sec. What must be the capacity, in scfm, of the air service to accomplish this work?

Solution.

Step 1. Determine the pressure:

$$P = \text{load/area} = 7060 \text{ lb}/78.54 \text{ in}^2 = 90 \text{ psi},$$

$$P \text{ (absolute)} = 90 + 14.7 = 104.7 \text{ psia}.$$

Step 2. Determine the flow rate:

a) $V_d = A_p \cdot S = 78.54 \text{ in}^2 \times 30 \text{ in.} = 2360 \text{ in}^3;$

b) $Q = V_d/t = 2360 \text{ in}^3/5 \text{ sec} = 472 \text{ in}^3/\text{sec}.$

Step 3. Calculate the compression ratio:

$$R_c = P_c/P_a = 104.7 \text{ psia}/14.7 \text{ psia} = 7.12 : 1.$$

Step 4. Assuming that an isothermal (constant temperature) condition exists, we obtain

$$Q_s = 472 \text{ in}^3/\text{sec} \times 7.12 = 3360 \text{ in}^3/\text{sec of free air}$$
$$= 1.945 \times 60 = 116.6 \text{ scfm}.$$

Thus the capacity of the air service must be 116.6 scfm.

Note that in the above example no consideration was given to any volumetric change of air as a function of pressure. It was treated as an incompressible fluid, so long as the load resistance remained constant. The compressibility of the fluid was recognized only when we converted from the flow rate of the fluid at operating pressure, 104.7 psia, to equivalent flow rate of free air, that is, air at standard conditions. This calculation is necessary when we relate the demand of the application to the capacity of the compressor system for delivering air.

Air Flow Losses in Pipes

We have discussed at some length the factors contributing to the loss of energy (head or pressure loss) in pipe flow involving liquid media. In treating this material, we did not make any provision for fluid compressibility, since liquids are generally considered incompressible at the normally encountered pressures. The same comment might be made regarding changes in temperature. We assume that an isothermal (constant temperature) condition exists, since the thermal expansion characteristics of a liquid are very small by comparison to a gas. Finally, since the pipe flow-loss techniques discussed are based on empirical solutions to the problems, variations due to pressure or temperature changes would be taken into account by the empirical data used.

The approach to compressible fluid flow systems is not quite so simple. In the first place, the volume of a gas varies inversely as the pressure and directly as the temperature. Thus, since the flow losses themselves result in a reduced pressure or a falling pressure gradient along the length of the pipe, they contribute to a change in specific volume as the gas moves through the pipe. We also know from thermodynamics that the compression of a gas is accompanied by the generation of heat, which must be dissipated to prevent a rise in pressure. And the expansion of a gas is accompanied by the release of heat. As in the case of the incompressible fluid, these effects introduce into the problem of flow-loss calculation variables which are not so easily handled.

There is an empirically derived formula which can be used in calculating approximate pressure drops due to the air flow in pipe systems. This is the Harris formula:

$$P_f = \frac{cL}{r} \cdot \frac{Q^2}{d^5}, \tag{17-4}$$

where P_f = pressure drop, psi,
L = length of pipe, ft,
r = compression ratio, pressure in pipe/atmospheric pressure,
Q = flow rate, ft^3/sec (free air),
d = actual internal pipe diameter.

For commercial pipe the coefficient is approximately

$$c = 0.1025/d^{0.31}.$$

Because of the lengthy calculations involved in obtaining the numerical solution of this and similar formulas, most fluid power engineers use precalculated and tabulated data to estimate the loss due to flow of air in pipes. This information is available in handbooks treating pneumatics, etc.

Table 17–1 summarizes some of the data calculated by means of the Harris formula. The value of the N-factor given in the table, when divided by the compression ratio, yields the pressure loss per 1000 ft of pipe.

EXAMPLE

Assume that we have 450 ft of $2\frac{1}{2}$-in. pipe through which free air passes at the rate of 575 ft^3/min. The discharge pressure of the compressor is 120 psi. What is the pressure drop in the line?

Solution.

Step 1. Using Table 17–1, we find that the N-factor of a $2\frac{1}{2}$-in. pipe handling 575 cfm of air is 76.7.

Step 2. We need to calculate the compression ratio. Since (a) the atmospheric pressure is 14.7 psia, it follows that (b) the final pressure is

$$120 + 14.7 = 134.7 \text{ psia.}$$

Hence (c) the compression ratio is

$$134.7/14.7 = 9.15 : 1.$$

Step 3. $76.7/9.15 = 8.4$ psi/1000 ft of pipe.

Step 4. $8.4 \times 450/1000 = 3.77$ psi drop for 450 feet of pipe.

Table 17–2 lists some calculated values of $d^{5.33}$ for several commercial pipe sizes. These data are used in the evaluation of the Harris formula for specific pipe systems.

EXAMPLE

Use the same parameters as for the example above. Using the Harris formula and substituting the data given, we obtain

$$p_f = \frac{0.1025L}{r} \cdot \frac{Q^2}{d^{5.31}} = \frac{0.1025 \times 450}{9.15} \cdot \frac{(575/60)^2}{124} = 3.74 \text{ psi},$$

which is in agreement with the value calculated above.

TABLE 17–1

FLOW FACTORS OF COMPRESSED AIR PASSING THROUGH STANDARD WEIGHT

Free air, cfm	*Nominal diameter, in.*								
	$\frac{1}{2}$	$\frac{3}{4}$	1	$1\frac{1}{4}$	$1\frac{1}{2}$	$1\frac{3}{4}$	2	$2\frac{1}{2}$	3
5	8.9	2.0	0.5	—	—	—	—	—	—
10	35.4	8.0	2.2	0.5	—	—	—	—	—
15	79.7	17.9	4.9	1.1	—	—	—	—	—
20	142	31.8	8.7	2.0	0.9	—	—	—	—
25	221	49.7	13.6	3.2	1.4	.7	—	—	—
30	318	71	19.6	4.5	2.0	1.1	—	—	—
35	434	97.5	26.6	6.2	2.7	1.4	—	—	—
40	567	127	34.8	8.1	3.6	1.9	—	—	—
45	716	161	44.0	10.2	4.5	2.4	1.2	—	—
50	885	199	54.4	12.6	5.6	2.9	1.5	—	—
60	—	286	78.3	18.2	8.0	4.2	2.2	—	—
70	—	390	106.6	24.7	10.9	5.7	2.9	1.1	—
80	—	510	139.2	32.3	14.3	7.5	3.8	1.5	—
90	—	645	176.2	40.9	18.1	9.5	4.8	1.9	—
100	—	796	217.4	50.5	22.3	11.7	6.0	2.3	—
110	—	963	263	61.1	27.0	14.1	7.2	2.8	—
120	—	—	318	72.7	32.2	16.8	8.6	3.3	—
130	—	—	369	85.3	37.8	19.7	10.1	3.9	1.2
140	—	—	426	98.9	43.8	22.9	11.7	4.6	1.4
150	—	—	490	113.6	50.3	26.3	13.4	5.2	1.6
160	—	—	570	120.3	57.2	29.9	15.3	5.9	1.9
170	—	—	628	145.8	64.6	33.7	17.6	6.7	2.1
180	—	—	705	163.3	72.6	37.9	19.4	7.5	2.4
190	—	—	785	177	80.7	42.2	21.5	8.4	2.6
200	—	—	870	202	89.4	46.7	23.9	9.3	2.9
220	—	—	—	244	108.2	56.5	28.9	11.3	3.5
240	—	—	—	291	128.7	67.3	34.4	13.4	4.2
260	—	—	—	341	151	79.0	40.3	15.7	4.9
280	—	—	—	395	175	91.6	46.8	18.2	5.7
300	—	—	—	454	201	105.1	53.7	20.9	6.6

Free air, cfm	2	$2\frac{1}{2}$	3	$3\frac{1}{2}$	4	$4\frac{1}{2}$	5	6	8	10
320	61.1	23.8	7.5	3.5	—	—	—	—	—	—
340	69.0	26.8	8.4	3.9	2.0	—	—	—	—	—
360	77.3	30.1	9.5	4.4	2.2	—	—	—	—	—
380	86.1	33.5	10.5	4.9	2.5	—	—	—	—	—
400	94.7	37.1	11.7	5.4	2.7	—	—	—	—	—
420	105.2	40.9	12.9	6.0	3.1	—	—	—	—	—
440	115.5	44.9	14.1	6.6	3.4	—	—	—	—	—

TABLE 17–1 (cont.)

STEEL PIPE. DATA BASED ON THE HARRIS FORMULA.

Free air, cfm	*Nominal diameter, in.*									
	2	$2\frac{1}{2}$	3	$3\frac{1}{2}$	4	$4\frac{1}{2}$	5	6	8	10
460	125.6	48.8	15.4	7.1	3.7	2.0	—	—	—	—
480	137.6	53.4	16.8	7.8	4.0	2.2	—	—	—	—
500	150.0	58.0	18.3	8.5	4.3	2.4	—	—	—	—
525	165.0	64.2	20.2	9.4	4.8	2.6	—	—	—	—
550	181.5	70.2	22.1	10.2	5.2	2.9	—	—	—	—
575	197	76.7	24.2	11.2	5.7	3.1	—	—	—	—
600	215	83.5	26.3	12.2	6.2	3.4	—	—	—	—
625	233	92.7	28.5	13.2	6.8	3.7	—	—	—	—
650	253	98.0	30.9	14.3	7.3	4.0	2.2	—	—	—
675	272	105.7	33.3	15.4	7.9	4.3	2.4	—	—	—
700	294	113.7	35.8	16.6	8.5	4.6	2.6	—	—	—
750	337	130.5	41.1	19.0	9.7	5.3	2.9	—	—	—
800	382	148.4	46.7	21.7	11.1	6.1	3.3	—	—	—
850	433	168	52.8	24.4	12.5	6.8	3.8	—	—	—
900	468	188	59.1	27.4	14.0	7.7	4.2	—	—	—
950	541	209.4	65.9	30.5	15.7	8.6	4.7	—	—	—
1000	600	232.0	73.0	33.8	17.3	9.5	5.2	1.9	—	—
1050	658	256	80.5	37.8	19.1	10.4	5.8	2.1	—	—
1100	723	280.6	88.4	40.9	21.0	11.5	6.3	2.4	—	—
1150	790	306.8	96.6	44.7	22.9	12.5	6.9	2.6	—	—
1200	850	344.0	105.2	48.8	25.0	13.7	7.5	3.3	—	—
1300	—	392.0	123.4	57.2	29.3	16.0	8.8	8.8	—	—
1400	—	—	—	66.3	33.9	18.6	10.2	3.8	—	—
1500	—	—	—	76.1	39.0	21.3	11.8	4.4	—	—
1600	—	—	—	86.6	44.3	24.2	13.4	5.1	—	—
1700	—	—	—	97.8	50.1	27.4	15.1	5.7	—	—
1800	—	—	—	110.0	56.1	30.7	16.9	6.4	—	—
1900	—	—	—	122	62.7	34.2	18.9	7.1	1.6	—
2000	—	—	—	135	69.3	37.9	20.9	7.8	1.8	—
2100	—	—	—	149	76.4	40.8	23.0	8.7	2.0	—
2200	—	—	—	166	83.6	45.8	25.3	9.5	2.2	—
2300	—	—	—	179	91.6	50.1	27.6	10.4	2.4	—
2400	—	—	—	195	99.8	54.6	30.1	11.3	2.6	—
2500	—	—	—	212	108.3	59.2	32.6	12.3	2.9	—
2600	—	—	—	229	117.2	64.9	35.3	13.3	3.1	—
2700	—	—	—	247	126	69.1	38.1	14.3	3.3	—
2800	—	—	—	265	136	74.3	41.0	15.4	3.6	—
2900	—	—	—	285	146	79.8	43.9	16.5	3.9	—
3000	—	—	—	305	156	85.2	47.0	17.7	4.1	—

TABLE 17–2

CALCULATED VALUES OF $d^{5.33}$ FOR USE WITH THE HARRIS FORMULA

Pipe of Nominal size, in.	$d^{5.33}$	*Pipe of Nominal size, in.*	$d^{5.33}$
$\frac{3}{8}$	0.0230	$2\frac{1}{2}$	124.0
$\frac{1}{2}$	0.0631	3	395.0
$\frac{3}{4}$	0.3561	$3\frac{1}{2}$	857.5
1	1.291	4	1,683.0
$1\frac{1}{4}$	5.572	5	5,627.0
$1\frac{1}{2}$	12.68	6	14,970.0
2	48.08		

The same technique can be used to calculate losses in branch lines. First we must calculate the loss up to the entrance to the branch line. Then we can use the new pressure as the initial pressure to the branch. The loss in the *run* can be calculated by using the Harris formula, as shown in the example. Any variable specified in the Harris formula can be calculated if the other variables are known.

Control with Orifices

When we discussed orifices and orifice-like devices in Chapters 8 and 9, we considered only incompressible fluids, i.e., liquids. The general form of the orifice flow equation was $Q = C_d A_v \sqrt{2gh}$. We found that values of C_d, the discharge coefficient, could be experimentally determined and would give fairly accurate results over a wide range of conditions.

In the case of compressible fluids, we run into the thermodynamic problems associated with the expansion of a gas as it is throttled (i.e., as it undergoes a pressure reduction) across an orifice, and the treatment becomes more complex than that for the liquid case.

As was pointed out in Chapter 9, most control devices for fluid power systems behave in an orifice-like manner. Thus, to be able to exercise control over an air system, we must be able to handle problems involving air flow through orifices and orifice-like devices.

It can be demonstrated that the mass-flow rate through an orifice is influenced by the downstream pressure when the pressure drop across the orifice is low enough for the velocity at the orifice throat to be *subsonic*. On the other hand, when the expansion ratio across the orifice is high enough to cause *sonic or supersonic velocities* at the orifice throat, the mass-flow rate is independent of downstream pressure.

This critical point is reached when the absolute pressure downstream of the orifice reaches 53% of the absolute pressure upstream. When this occurs, that is, when $P_{2_a} = 0.53P_{1_a}$, the throat velocity is equal to the speed of sound in that air. When $P_{2_a} > 0.53P_{1_a}$, the following flow equation is applicable:

$$Q_m = \frac{0.595C \cdot D^2 \cdot P_2}{T_1 \cdot \sqrt{(n - 1/n)}} \cdot \sqrt{\frac{P_1}{P_2}} \times \sqrt{\left(\frac{P_1}{P_2}\right)^{(n-1)/n} \cdot \left[\left(\frac{P_1}{P_2}\right)^{(n-1)/n} - 1\right]} \tag{17-5}$$

where Q_m = lb air/min

C = discharge coefficient (about 0.65 for sharp-edged orifice, 0.96 to 0.99 for round-edged orifices),

D = diameter of orifice throat, in.,

P_1 = upstream pressure,

P_2 = downstream pressure,

T_1 = upstream temperature, °R (absolute temperature),

n = ratio of specific heats of air at constant pressure to constant volume ($n = 1.406$ for dry air).

When $P_{2_a} < 0.53P_{1_a}$, the following expression is applicable:

$$Q_m = 0.5303 \frac{a \cdot c \cdot P_{1_a}}{1\sqrt{T_1}}, \tag{17-6}$$

where a = area of orifice, in^2. It is evident that the air case is more complex than the liquid case. As in the instance of calculating losses in pipe lines, data on air flow through orifices have been reduced to tabular form to facilitate practical calculations. Table 17–3 summarizes some of these data for air flow through orifices.

Air Control Valves

Because of the complexities just discussed, most manufacturers of air control valves have adopted one or the other of several systems to rate their product; i.e., they use for this purpose various factors, called C_v-factor, F-factor, K-factor, etc. In essence this approach is based on the concept that the "factor", when multiplied by the pressure drop (or some function of pressure drop) across the valve, will give the mass-flow rate of air through the valve. A detailed discussion of these methods is beyond the scope of this text.

TABLE 17–3 CHART OF FREE AIR AT CFM THROUGH ORIFICE*

Supply pressure, psi	*Orifice size, diameter*																	
	$\frac{1}{32}$	$\frac{1}{16}$	$\frac{3}{32}$	$\frac{1}{8}$	$\frac{5}{32}$	$\frac{3}{16}$	$\frac{7}{32}$	$\frac{1}{4}$	$\frac{9}{32}$	$\frac{5}{16}$	$\frac{11}{32}$	$\frac{3}{8}$	$\frac{13}{32}$	$\frac{7}{16}$	$\frac{15}{32}$	$\frac{1}{2}$	$\frac{3}{4}$	1
2	0.17	0.62	1.44	2.55	3.93	5.74	7.71	10.2	13.0	15.9	19.2	23.2	27.1	31.5	35.9	41.0	92.0	164.0
5	0.25	0.94	2.19	3.77	5.95	8.70	11.6	15.4	19.6	24.0	29.0	35.0	40.9	47.6	54.3	62.0	139.0	248.0
10	0.39	1.48	3.41	6.10	9.31	13.6	18.2	24.1	30.6	37.5	45.4	54.8	64.0	74.5	85.0	97.0	218.5	388.0
15	0.41	1.62	3.63	6.44	10.0	15.4	19.6	25.8	32.6	40.0	48.7	58.2	68.1	78.8	90.7	103.1	227.9	412.5
20	0.49	1.90	4.28	7.59	11.8	17.0	23.1	30.4	38.4	47.2	57.4	68.6	80.2	92.9	106.9	121.5	268.7	486.3
25	0.56	2.18	4.93	8.74	13.6	19.6	26.6	35.0	44.2	54.3	66.1	79.0	92.4	107.0	122.1	140.0	309.4	560.0
30	0.63	2.47	5.58	9.89	15.4	22.2	30.4	39.6	50.1	61.5	74.8	89.4	104.6	121.1	138.3	158.4	350.2	633.8
35	0.71	2.76	6.23	11.0	17.2	4.8	34.0	44.2	55.9	68.6	83.5	99.8	116.8	135.1	154.5	176.8	390.9	707.6
40	0.78	3.05	6.88	12.2	19.0	27.3	37.5	48.8	61.7	75.8	92.2	110.2	128.9	149.2	170.7	195.3	431.7	781.4
45	0.85	3.34	7.53	13.3	20.8	29.9	41.0	53.4	66.6	82.9	100.9	120.6	141.1	163.3	186.9	213.8	472.5	855.1
50	0.92	3.62	8.17	14.5	22.6	32.5	44.6	57.0	72.4	90.1	109.6	131.0	153.3	177.4	203.1	232.2	513.2	928.9
55	1.00	3.91	8.82	15.6	24.4	35.1	48.1	62.7	78.2	97.3	118.3	141.4	165.4	191.5	219.3	250.5	554.0	1002.7
60	1.07	4.20	9.47	16.8	26.2	37.7	51.6	67.2	84.1	104.4	127.0	151.8	177.6	205.6	235.5	268.9	594.7	1075.4
65	1.15	4.49	10.1	17.9	27.9	40.3	55.2	71.8	89.9	111.7	135.7	162.2	189.8	219.7	251.7	287.3	634.8	1149.2
70	1.21	4.77	10.8	19.1	29.7	42.8	58.8	76.4	95.7	118.8	144.4	172.6	202.0	233.8	267.9	305.8	675.7	1223.0
75	1.30	5.06	11.4	20.2	31.5	45.4	62.3	81.0	105.5	126.0	153.1	183.0	214.1	247.9	284.1	324.2	716.4	1296.7
80	1.37	5.35	12.1	21.4	33.3	48.0	65.8	85.6	107.4	133.1	161.8	193.4	226.3	262.0	300.3	342.6	757.2	1370.5
85	1.44	5.64	12.7	22.5	35.1	50.6	69.4	90.3	113.2	140.3	170.5	203.8	238.5	276.1	316.5	361.1	797.9	1444.3
90	1.52	5.92	13.4	23.7	36.9	53.2	72.9	94.8	119.0	147.5	179.2	214.2	250.6	290.1	332.7	379.4	838.7	1518.1
95	1.59	6.21	14.0	24.8	38.7	55.7	76.5	99.4	124.9	154.6	188.0	224.6	262.8	304.2	358.9	398.0	879.4	2591.8
100	1.66	6.50	14.7	26.0	40.5	58.3	80.0	104.6	130.7	161.8	196.7	235.0	275.0	318.3	375.1	416.2	920.2	1664.2
125	2.03	7.94	17.9	31.7	49.5	71.4	97.7	127.1	159.8	197.5	240.2	287.0	335.8	388.8	464.8	508.3	1124.0	2033.2
150	2.40	9.28	21.2	37.5	58.4	84.4	115.4	150.1	189.0	233.3	283.7	339.0	396.7	459.2	545.2	600.4	1327.8	2401.5

* For pressure below atmospheric, use standard adiabatic formula.

Accumulators

Figure 17–2 illustrates a device which makes use of both incompressible and compressible fluids. It is called an *accumulator,* because it accumulates, or stores, energy. An interface is placed between the liquid and gas sides of the device. In some designs the interface is a rubber bladder, as in Fig. 17–2; in others, a piston is used. The interface is introduced to separate the two fluid phases to prevent the dissolving of the gas in the liquid medium. Initially, the accumulator is "precharged" by the introduction of gas (usually nitrogen) under pressure. (See Fig. 17–3a.) When the system is in operation, liquid is pumped into the accumulator by the pump. Since the operating pressure of the system is greater than the precharge pressure in the accumulator, the gas is compressed to the new pressure, P_s. Increasing the pressure on the gas reduces its volume, so that liquid, usually oil in a fluid power system, flows in to fill the void in the accumulator. (Fig. 17–3b). If, at any time, the system pressure drops below the pressure of the gas in the accumulator, the gas reexpands to match the reduced pressure. As a result, some of the liquid is forced out of the accumulator and back into the system. Thus an accumulator can be used to store energy over a long period of time and release it, as a supplement to the pump, over short time intervals. Accumulators, due to their ability to absorb energy, are also used to reduce shock in hydraulic systems.

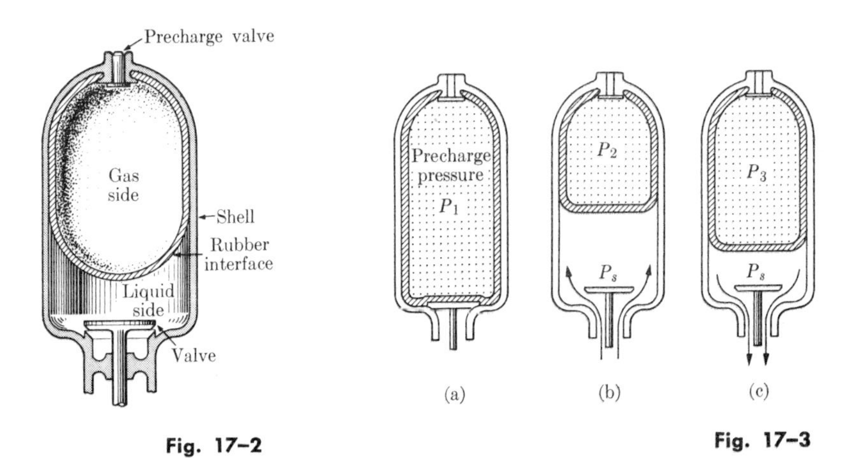

Fig. 17–2

Fig. 17–3

If we calculated the volume-pressure relationships in an accumulator strictly according to Boyle's law, we would be presupposing the existence of isothermal (constant temperature) conditions. Since we know that this is not exactly true in practice, because heat is generated by the compression of a gas, we would find our answers to be slightly in error.

Boyle's law can be written in the form

$$P_1V_1^n = P_2V_2^n = K. \qquad (17\text{–}7)$$

For isothermal processes, $n = 1$; for adiabatic processes $n = 1.4$. In practice, the process followed is *polytropic*, yielding $n = 1.25$.

EXAMPLE

Determine the size of an accumulator (volume) capable of supplying 250 in³ of oil with a pressure differential from 3000 psi (system) to 2000 psi (system). Assume a precharge pressure of 1500 psi.

Isothermal Solution.

Step 1. $P_1 = 3514.7$ psia, $P_2 = 2014.7$ psia, $P_3 = 3014.7$ psia.

Step 2. $V_1 =$ the required accumulator size,

$$V_2 = V_3 + 250 \text{ in}^3,$$
$$V_3 = V_2 - 250 \text{ in}^3.$$

Step 3. We determine V_3, using Boyle's law, $p_2V_2 = p_3V_3$. Substituting the values obtained in step 2, we have

$$2014.7\ (V_3 + 250) = 3014.7V_3,$$
$$1000\ V_3 = 503{,}000,$$
$$V_3 = 503 \text{ in}^3.$$

Step 4. $V_2 = V_3 + 250 = 753$ in³. We now have all the quantities needed to determine V_1.

Step 5. $P_1V_1 = P_2V_2$. Hence

$$1514.7\ V_1 = 2014.7 \times 753,$$
$$V_1 = 1000 \text{ in}^3.$$

Adiabatic Solution.

Step 1. $P_1V_1^{1.4} = P_2V_2^{1.4} = P_3V_3^{1.4}$.

Step 2.

$$V_2 = (P_3/P_2)^{0.714},$$
$$V_3 = (3014.7/2014.7)^{0.714}(V_2 - 250) = 1.33\ (V_2 - 250),$$
$$V_2 = 1000 \text{ in}^3.$$

Step 3. $V_1 = (2014.7/1514.7)^{0.714} \times 1000 = 1230 \text{ in}^3.$

Note the percentage difference in size determined by the two methods of calculation:

$$\frac{1230 - 1000}{1230} = 18.7\%.$$

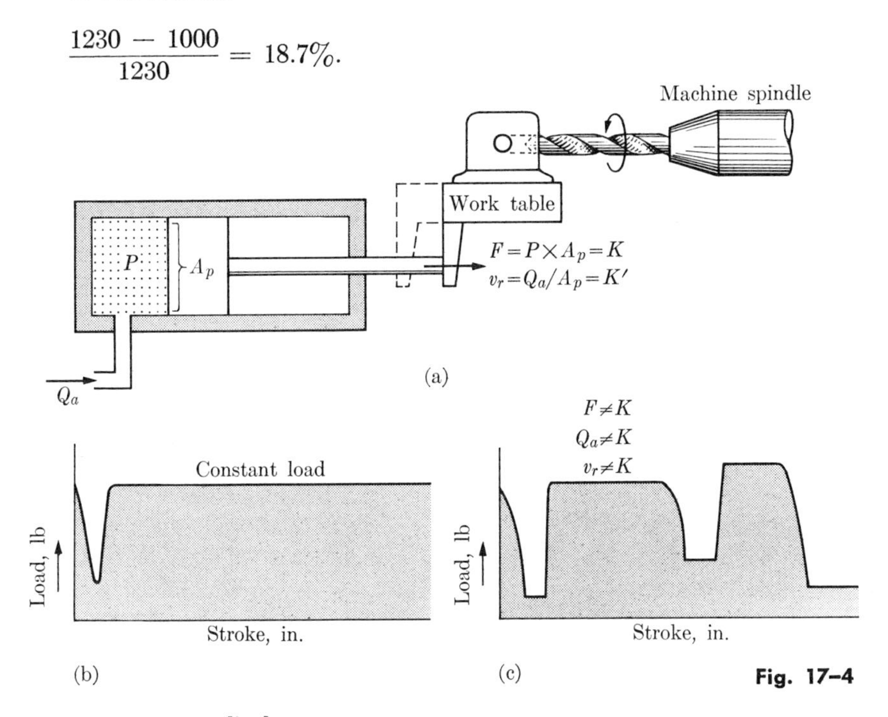

Fig. 17–4

Pneumatic cylinders

From earlier chapters we recall that in calculations involving cylinders a constant load reaction on the piston rod was assumed. Thus the pressure was easily calculated to be $P = F/A$. Even if the load reaction did vary, only the pressure would change, for we were concerned only with incompressible fluids. This situation does not hold for compressible fluids. Figure 17–4 illustrates what happens with pneumatic systems when the load reaction changes. Figure 17–4(a) shows a typical cylinder as it might be used in a machine tool application. The purpose of the cylinder is to push, or feed, the work piece into the cutting tool. So long as the load remains constant, the pressure will be constant (Fig. 17–4b). Thus the gas will behave like an incompressible fluid.

Suppose that the load varies, as indicated in Fig. 17–4(c). As the pressure varies, the gas volume will change accordingly. Thus the gas in the cylinder will undergo expansion or compression as a function of load variation. Under these conditions it is impossible to maintain a constant piston rod velocity; that is, the feed rate will change. For this reason pneumatic cylinders are rarely used in applications where constant velocities must be maintained, unless auxiliary control devices are used.

IMPORTANT TERMS

Charles' law is a gas law which relates pressure, temperature, and volume of a gas as a function of temperature.

Boyle's law is a gas law relating pressure and volume, with temperature held constant.

General gas law relates temperature, volume, and pressure as a function of the universal gas constant.

Compressor is a mechanical device for transferring energy to compressible fluids.

Pneumatic fluid power system is an energy transfer system in which the transfer medium is a compressible fluid, usually air.

Harris formula is an empirical formula for the calculation of flow losses of air in pipes.

Compression ratio is the ratio of the absolute pressure of a gas at some pressure level to the absolute pressure of gas under standard conditions.

Standard conditions exist when the atmospheric pressure of air is taken at sea level at 68°F.

Critical pressure in orifice flow is the amount of downstream absolute pressure which is equal to 53% of the absolute upstream pressure.

Accumulator is a gas-liquid device used in fluid power systems to store energy.

Pneumatic cylinder is a linear force-producing device, used in fluid power systems, which utilizes pressurized gas to generate a force.

PROBLEMS

17–1 Standard air enters a horizontal pipe 6 in. in diameter with a velocity of 370 ft/sec. The exit velocity is 45 fps. Neglecting friction, calculate exit pressure and temperature.

17–2 Air enters an 8-in. pipe with an inlet pressure of 110 psi and a temperature of 90°F. The inlet velocity is 200 fps. Exit conditions include pressure of 70 psi and a temperature of 110°F. Calculate the friction opposing motion.

17–3 A stream of oats and air is blown through a combine (an agricultural machine used in harvesting grain). A sheet metal duct is used to guide the flowing mixture. There is a bend in the duct which deflects the stream through an angle of 120°. The duct has a uniform cross-sectional area of 12 in. × 12 in. The average inlet velocity is 10 fps, and the average specific weight of the mixture is 52 lb/ft^3. What force does the air-oats mixture exert on the bend in the duct?

17–4 A 2-in. pipe 500 feet long handles free air at a rate of 500 ft^3/min. The inlet pressure is 110 psi. What is the outlet pressure?

17–5 The inlet pressure to a 1000-ft run of $1\frac{1}{2}$-in. pipe is 100 psi. If the outlet pressure is 95 psi, what is the flow rate of air in scfm?

17–6 Air flows through a pipe 2 in. in diameter at the rate of 600 ft^3/min. The inlet pressure is 120 psi and the outlet pressure is 110 psi. How long is the pipe run?

17–7 Given a $\frac{1}{4}$-in. sharp-edged orifice used to control air flow. The upstream pressure is 120 psi, and the downstream pressure is 50 psi. The temperature at the orifice inlet is 95°F. Calculate, in scfm, the flow rate across the orifice.

17–8 Given a $\frac{1}{32}$-in. round-edged orifice. Air is fed in at a temperature of 110°F and a pressure of 100 psi. The downstream pressure is 40 psi. Calculate the flow rate of air in scfm.

17–9 Air approaches a sharp-edged orifice $\frac{1}{2}$ in. in diameter at 100°F and 95 psi. Downstream pressure is 55 psi. Calculate the flow rate in scfm.

17–10 Assume that we have a transfer cylinder which must raise a 100-lb casting a distance of 12 in. (Fig. 17–5). With an air pressure of 90 psi (at the cylinder), what size cylinder must we use? How much power is consumed in raising the casting?

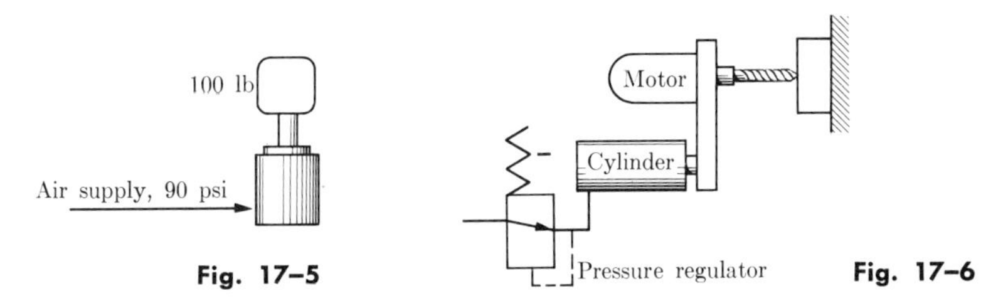

Fig. 17–5 **Fig. 17–6**

17–11 Figure 17–6 shows an air-fed cylinder powering a drill-feed unit. If it takes a force of 282 lb to push the drill through the casting, and the air cylinder is 90% efficient, what must the regulator setting be for a 2-in. bore cylinder?

17–12 In a transfer application, as shown in Fig. 17–7, a 500-lb load is to be moved over a set of ways. It has been determined that the coefficient

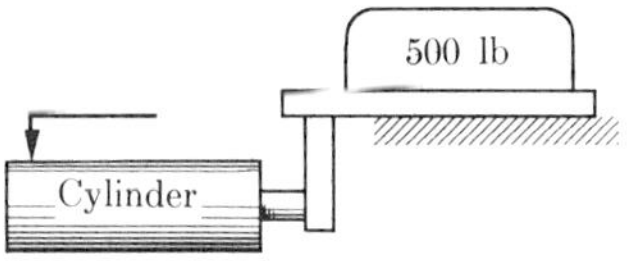

Fig. 17–7

of friction between the table and ways is 0.15. Tests have also shown that the cylinder has a static friction load equal to 10% of the maximum rated load. Also, the cylinder is 90% efficient during operation. With air pressure of 65 psi available, what size should the cylinder be?

17–13 In Problem 17–11, it takes a force of 100 lb to withdraw the drill from the casting. If the cylinder has a rod 1 in. in diameter, will the regulator setting, as calculated in Problem 17–11, be satisfactory?

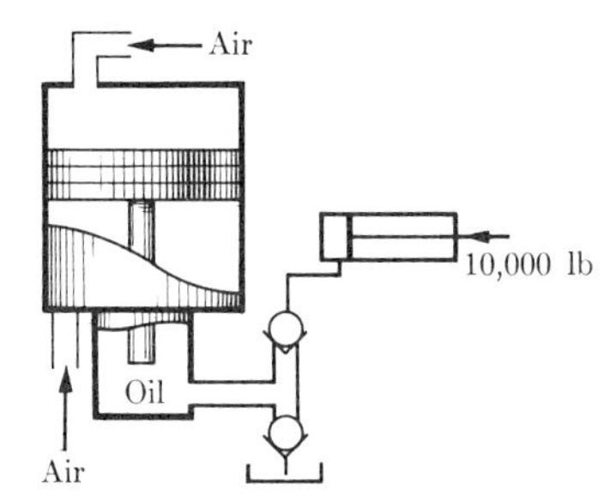

Fig. 17–8

17–14 An air-hydraulic intensifier is connected to a 2 in^2 (cross section) hydraulic cylinder with a 10,000-lb load on it. The intensifier rod has an area of $\frac{1}{2}$ in^2. The intensifier is 92.7% efficient. If the piston is 7 in. in diameter, what air pressure is required to drive it? If the stroke of the intensifier is 2 in., what is the displacement of the hydraulic pump? What is the displacement of the blank end of the air cylinder? What is the displacement of the rod end? If the intensifier makes 1 stroke/sec, what is the flow rate of oil from it? What is the free-air consumption, in cfm?

17–15 If the stroke in the cylinder of Fig. 17–7 is 30 in. and the load must be moved in 5 sec, what will be the speed (velocity) of the piston rod? the volume of the cylinder? the flow rate at 65 psi? the volume of free air required? the consumption of free air?

17–16 What is the compression ratio in the cylinder of Fig. 17–7? What is the approximate size of the orifice needed in the valve to give this flow rate?

17–17 In Problem 17–12, what would be the receiver pressure if air were supplied to the cylinder through 2000 ft of $\frac{1}{2}$-in. pipe?

17–18 Assume that in Fig. 17–9 $P_1 - P_2$ (i.e., the pressure drop across the orifice) is constant. If the operating pressure level is 50 psig, what is the flow rate of free air in cfm across a $\frac{1}{8}$-in. orifice? If the operating

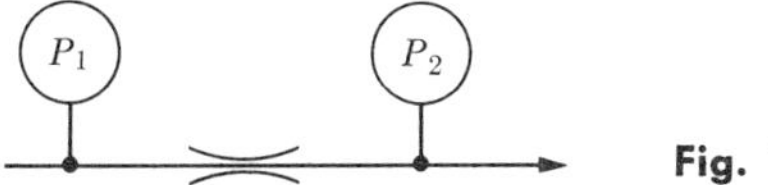

Fig. 17-9

pressure is changed to 70 psig and the pressure drop is 10 psig, what will the new flow rate be? How does this result vary from what is observed with an hydraulic fluid?

17-19 We have an arm 3 ft long supporting a 100-lb weight. An air cylinder is pivoted a distance of 1 ft from the arm pivot. (a) Determine the varying load conditions on the cylinder. (b) What will happen to the cylinder speed as the arm swings? (c) What will happen to the air consumption? (See Fig. 17-10.)

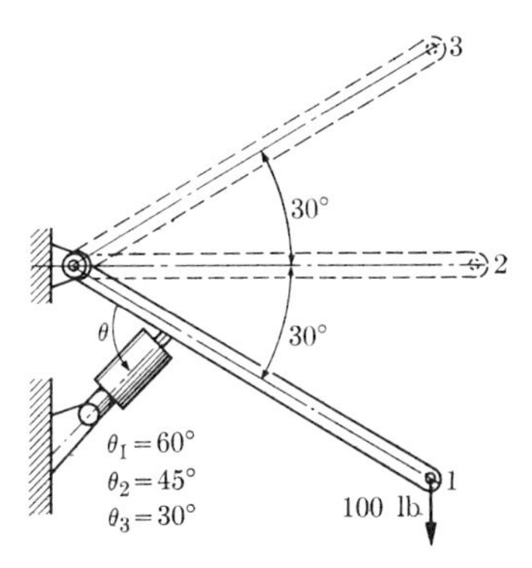

Fig. 17-10

17-20 An accumulator is to supply 400 in^3 of oil with a maximum pressure of 3000 psi and a minimum system pressure of 1800 psi. The nitrogen precharge pressure is 1200 psi. Calculate the size of the accumulator. (See Fig. 17-11.)

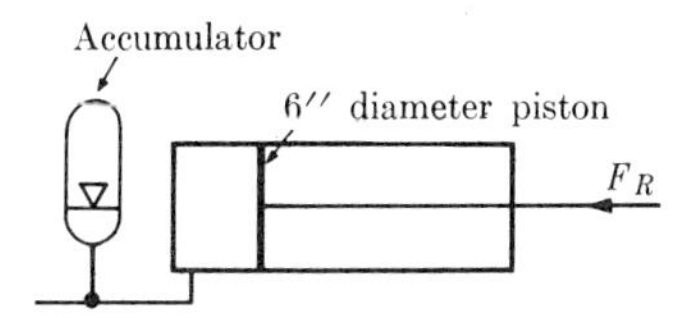

Fig. 17-11

17-21 Given that the accumulator of Problem 17-20 is driving the cylinder in Fig. 17-11, determine the variation in load which the cylinder would be capable of handling over its entire stroke. What would be the stroke of the cylinder if it used the entire output of the accumulator?

17-22 In the system of Problem 17-20 the load reaches a magnitude of 62,500 lb. What is the stroke length at which this load would just balance the pressure in the accumulator? Could the cylinder stroke extend beyond this point? Explain.

17–23 An accumulator must supply 3 gal of fluid to a system. The maximum system pressure is expected to be 2500 psi, and the minimum allowable pressure is 1500 psi. The nitrogen precharge pressure is 800 psi. Calculate the required accumulator size. Assuming adequate structural strength, determine what would happen if the actual system pressure went to 3000 psi. Under these conditions, what would be the pressure of the nitrogen in the accumulator?

*17–24 From what you know of pressure, load, and density relationships and the gas laws, make the necessary calculations and a plot of the variation of pressure and density of air as a function of the altitude above the earth's surface.

CHAPTER 18

Applications

Energy Transfer Devices, Input

The mechanical devices which we know as *pumps* are actually devices for the transfer of energy from some source, called the prime mover, to the fluid in a system. In industrial applications, with which we are concerned in this text, this energy transfer serves one of two basic reasons:

1) to move or transport the fluid, as in a process plant, etc.; or

2) to transfer energy to a mechanical load, as in a fluid power system.

Pumps

There are two basic modes in which energy can be transferred:

1) by means of kinetic energy, where KE $= \frac{1}{2}Mv^2$; or

2) by means of potential energy, where PE $= W(P_2 - P_1)/\gamma$.

In the first case, energy is transferred by increasing the velocity of the fluid. In the second case, it is transferred by raising the pressure.

Hydrodynamic devices

In industry, particularily in fluid power applications, the kinetic-energy transfer devices are called *hydrodynamic* pumps. The potential-energy transfer pumps are called *hydrostatic* pumps. The latter term is actually a misnomer, because the fluid does flow and is not static. However, the increase in potential energy of the fluid after it passes through the pump is evidenced by an increase in hydrostatic pressure, and thus we derive the name.

Typical hydrodynamic devices encountered in industry are the centrifugal pump, turbine pump, fluid coupling, and torque converter. It is interesting to note that even though energy in a hydrodynamic device is transferred via kinetic energy, at some point in the device a diffuser section (see Chapter 12) must be employed to convert some of the kinetic energy to potential energy (pressure) in order to provide for line losses, lift, etc.

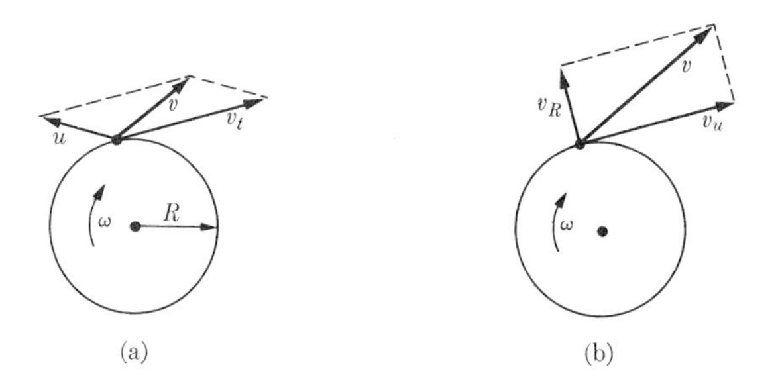

Fig. 18–1

Figure 18–1 illustrates a typical impeller or "wheel" of a centrifugal pump. The vector diagrams on the periphery of the impeller illustrate velocity relationships. In Fig. 18–1(a), the vector v_t represents the velocity of a point on the impeller. If R is the radius of the impeller and ω is the angular velocity, $v_t = R \cdot \omega$. The vector u is the velocity of the fluid relative to the impeller; v, the vector sum of u and v_t, is the absolute velocity of the fluid.

In Fig. 18–1(b), the components of v are shown: v_R is the radial component, and v_u is the tangential component of the fluid velocity. From our discussion in Chapters 11 and 12, we recall that momentum is defined as the product of mass and velocity, or mv. We also showed that the product of force and the time during which it acts is equal to the change in momentum, or

$$F \cdot t = m(v_2 - v_1).$$

Dividing through by t, we get

$$F = \frac{m(v_2 - v_1)}{t}.$$

This result tells us that the force acting on a body is equal to the change in momentum per unit time. In differential equation form, the above expression becomes

$$F = \frac{d}{dt}(mv).$$

This tells us that the force is equal to the time rate of change of momentum. If we multiply both sides of the equation by R and use the tangential velocity notation v_u, we get

$$FR = \frac{d}{dt}(mRv_u).$$

The product mRv_u is the angular momentum, and the expression now tells us that the torque is equal to the time rate of change of angular momentum. It can be shown that the rotor of such a pump acts on the fluid to change its angular momentum: $m(R_2v_{u_2} - R_1v_{u_1})$. Thus

$$T = M_1(R_2v_{u_2} - R_1v_{u_1}),$$

where M_1 is the mass flow rate. The torque is positive for a pump or compressor and negative for a turbine runner.

The expression for power or rate of doing work is $P = T\omega$. From the expression for torque in the development above, power could be expressed as

$$P = M_1(R_2v_{u_2} - R_1v_{u_1})\omega.$$

The discharge flow rate of a hydrodynamic pump of the type under discussion is proportional to the radial velocity component, v_R, and the runner exit area. The impeller exit area would be equal to the product of the circumference of the impeller and the width of the opening:

$$A_e = \pi D_0 \cdot b.$$

A factor frequently mentioned in evaluations of hydrodynamic devices is the *specific speed:*

$$N_s = \frac{N\sqrt{\text{gal/min}}}{h^{3/4}},$$

where N is measured in rpm, and h is the pump head per stage of the pump. The specific speed gives an indication of the pump head, speed, and flow-rate characteristics that are possible and, perhaps, desirable.

Hydrodynamic pumps are seldom found in fluid power applications, but are most often used in fluid transfer or materials handling applications. This separation of application areas is necessary for several reasons: (1) high enough pressures cannot be generated in one stage of the pump;

(2) the discharge flow rates vary with the change in head; and (3) the pump may "stall." Here stalling means that some limiting pressure head at the discharge port can be reached beyond which the pump cannot go. The stalling occurs because the fluid is not forced out into the system mechanically but depends for its flow on the change in velocity across a rotating impeller; therefore the impeller will shear the fluid at its periphery. Under these conditions, all the energy input to the pump is converted into head. This stalling condition is inadmissible in a fluid power system. On the other hand, it has some advantages in power transmission systems using hydrodynamic devices, such as torque converters and fluid couplings.

Hydrostatic devices

Because power transmission is the prime function of a fluid power system, it is desirable to maintain a steady, controllable rate of energy flow through the system. Figure 18–2 illustrates the difference between a mechanical transmission system, such as a belt drive (part a), and a fluid power transmission system (part b). In a mechanical transmission system, power is transmitted by an input pulley to an output pulley by means of the tension in the belt. In Fig. 18–2(b), a pump, instead of a pulley, is the input device; pipes, instead of a belt, transmit energy; and a fluid motor, instead of a second pulley, is the output device. The flow of fluid which moves from the pump through the pipe to the motor, where the energy is extracted, and then back to the pump is analogous to the flow of energy which travels from one pulley through the belt to the other pulley. We may consider the pressure on the fluid in the pipe to be equivalent to the tension in the belt.

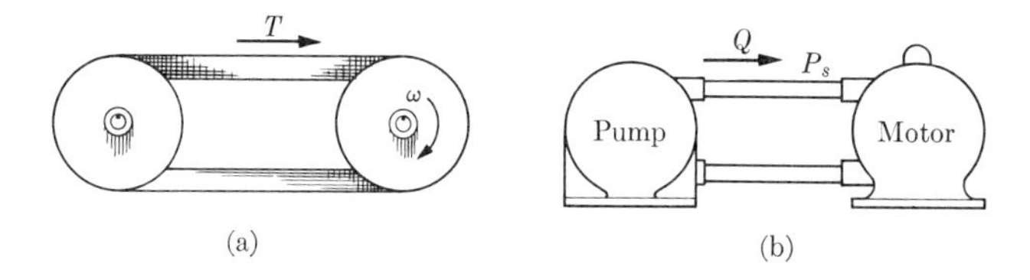

Fig. 18–2

From this qualitative description of a fluid power system, we can see that we must have a continuous flow of fluid at some pressure in order to transmit power. If the pump stopped forcing fluid into the pipe at some pressure, the system would stop functioning. For this reason, we use *positive displacement pumps*, rather than hydrodynamic pumps, in fluid power work.

Figure 18–3 illustrates the action of a positive displacement pump from the viewpoint of the external circuit. A PD pump will eject a relatively fixed quantity of fluid into the system for every cycle of the pump. This fixed quantity is called *displacement* and is a finite volume, not a rate.

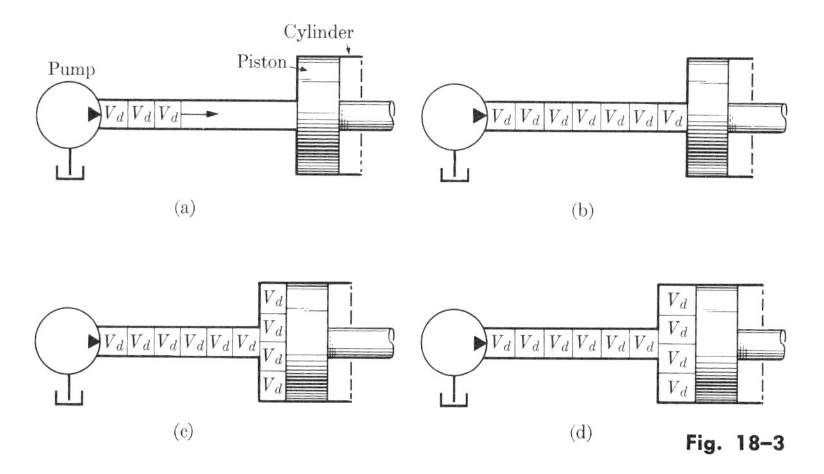

Fig. 18–3

It is illustrated, in Fig. 18–3(a), by using the symbol for a fluid power pump and indicating that slugs of fluid, V_d, are ejected into the pipe by the pump. When the pipe is full, as in Fig. 18–3(b), the next slug of fluid will cause the cylinder to push on the mechanical load. Neglecting line losses, we find that the pressure developed by the pump will be a direct function of the load resistance distributed across the face of the piston in the cylinder. Thus we see that positive displacement pumps do *not* pump pressure; they transfer fluid into a system against a load resistance. The latter is responsible for the generation of pressure; no load, no pressure. As the PD pump continues to transfer slugs of fluid into the line, the cylinder piston is pushed out farther, and the work is accomplished. All this occurs at very high speed, of course, so that the action appears to be a steady flow rather than a series of jerks and starts.

The design of fluid power pumps, or PD pumps, must provide for the intake of fluid through an inlet or suction port; its transport through the pumping mechanism in such a way that at some point in the pumping cycle the fluid slug is momentarily isolated from the outside world; sufficient energy transfer to the fluid to ensure that the pressure will be raised to that required by the particular load; and forcible (positive) ejection of the fluid slug into the system. You cannot stall a positive displacement pump, for it will continue to deliver fluid against a load resistance until either the prime mover stalls or something breaks.

The three most commonly used types of PD pumps are gear, vane, and piston. A typical gear pump design is illustrated in Fig. 18–4. Two or more meshing spur gears are placed within an enclosure, which completely surrounds the gears and isolates them from the outside environment. An inlet port is placed so as to connect with the gears on one side of the pump. A second port provides for outlet or discharge of the fluid.

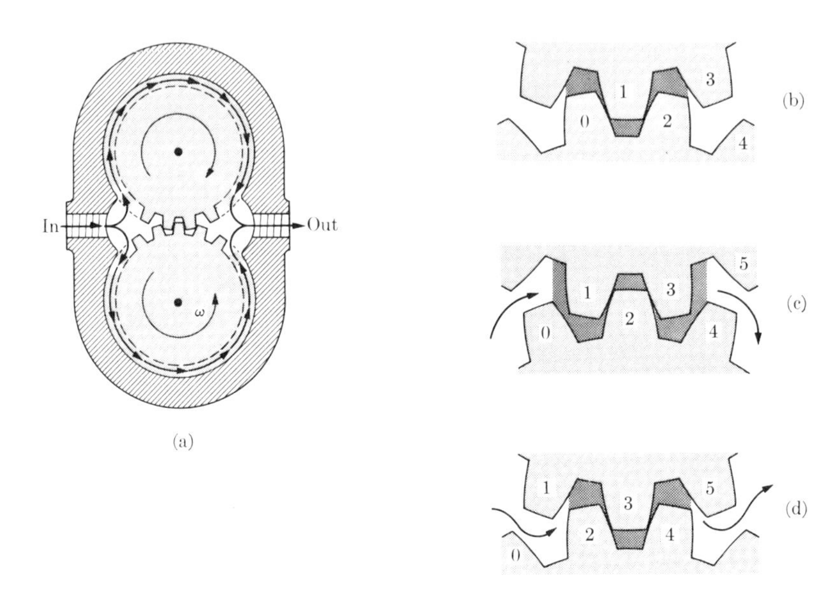
In
Out
ω
(a)
1
3
0
2
4
(b)
5
1
3
2
0
4
(c)
1
5
3
2
4
0
(d)

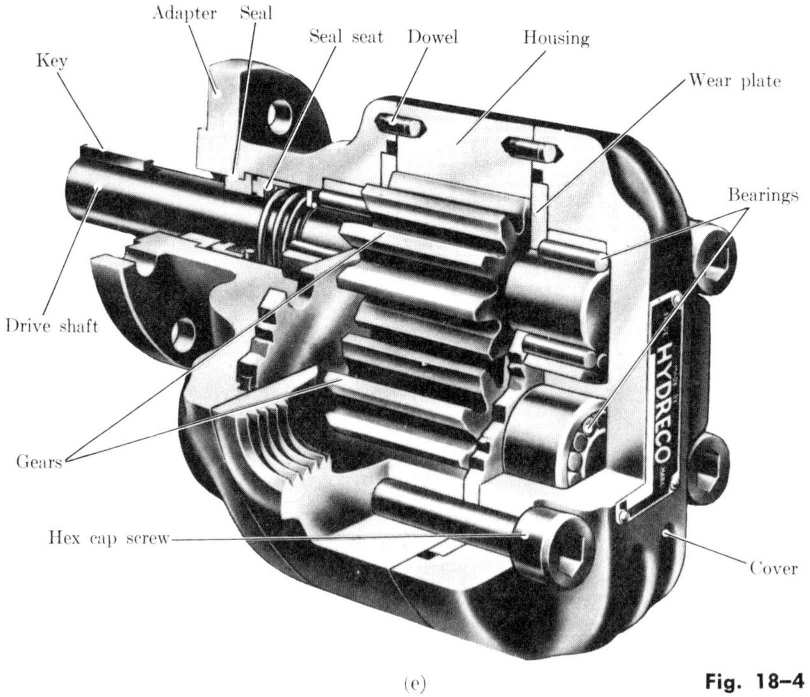
Adapter
Seal
Seal seat
Dowel
Housing
Key
Wear plate
Bearings
Drive shaft
HYDRECO
Gears
Hex cap screw
Cover
(e)

Fig. 18–4

As the gears rotate, the teeth come out of mesh on one side and go into mesh on the other. On the side where they come out of mesh the volume expands, bringing about a reduction in pressure which causes the fluid to be pushed into the void by atmospheric pressure. This process is usually called "suction." The term, however, is a misnomer. On the side where the teeth go into mesh the volume decreases between mating teeth. The decreasing volume forcibly ejects the fluid from the tooth spaces and discharges it into the outlet port of the pump. Because all the pumping action occurs in the region where the gears mesh, the fluid is picked up in the spaces between the teeth and carried around the periphery of the gears.

The theoretical output from a gear pump can be closely approximated from the following formula:

$$V_d = \frac{\pi}{4}(D^2 - d^2)w, \tag{18–1}$$

where V_d = displacement of the pump, in^3/rev,
D = outer diameter of the gear,
d = gear root diameter,
w = width of gear.

Flow rate is $Q = V_d \cdot N$, where N is given in rpm.

The second major class of positive displacement pumps used in fluid power applications is the vane pump, illustrated in Fig. 18–5. This type of pump uses a rotor carried on a shaft. The rotor has slots machined radially into it around its periphery. A cam ring is placed around the rotor in such a

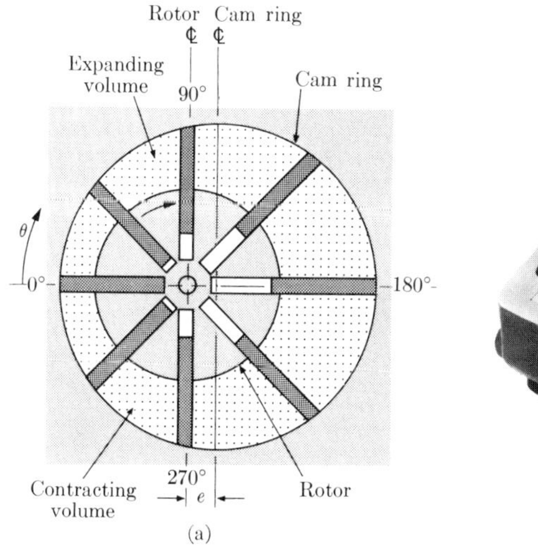

(a)

(b)

Fig. 18–5

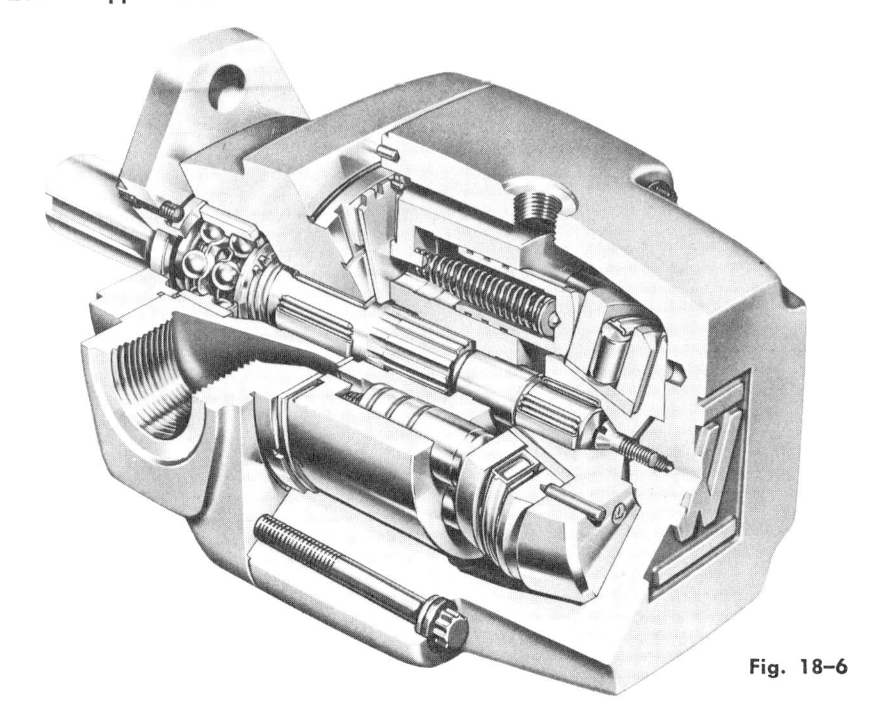

Fig. 18–6

position that the centerline of the cam ring is eccentric to the centerline of the rotor shaft. Starting at $\theta = 0°$, where the rotor and cam ring are closest together, as indicated in Fig. 18–5, the volume expands as the vanes rotate toward the 180° position. The centrifugal force keeps the vanes out against the surface of the cam ring. During the first 180° of rotation of the shaft-rotor assembly, the resulting volume expansion causes the reduction in pressure associated with the "suction" or induction process. Fluid flows in to fill the void through porting designed into the pump. As the vanes rotate through the second 180°, the surface of the cam ring pushes them back into their respective slots and the volume trapped between the vanes, rotor surface, and cam surface is reduced, thus forcibly ejecting the trapped fluid through suitable ports. We see, then, that the conditions of positive displacement are satisfied.

A simplified formula for the displacement of a vane pump is

$$V_d = 2\pi e\, Dw \text{ in}^3/\text{rev}, \tag{18–2}$$

where e = eccentricity,
D = cam ring diameter,
w = width of rotor elements,

or $Q = V_d N$ in^3/min, where N is given in rpm.

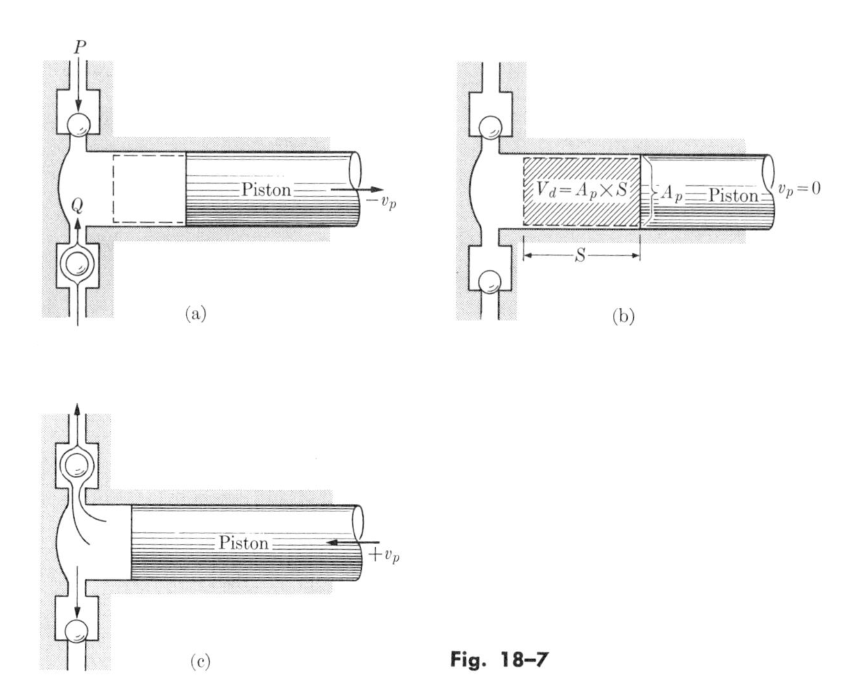

Fig. 18–7

The third class of pumps used in fluid power applications is the piston pump. A typical design is shown in Fig. 18–6. Piston pumps utilize several cylindrical pumping elements, or pistons, placed in closely fitting bores in a cylinder barrel. Various types of mechanisms are employed to make the pistons reciprocate within the bore. When the piston is withdrawn from the bore, the expanding volume within the cylinder causes the induction of fluid through suitable valving (Fig. 18–7a). As soon as the piston reaches the end of the "suction" stroke, the inlet valve closes. The direction of motion of the piston is then reversed, as indicated in Fig. 18–7(b). As a result, the fluid is pushed out ahead of the piston and through a discharge valve into the system, as in Fig. 18–7(c). The displacement volume of one piston is

$$V_d = A_p \cdot S,$$

where A_p is the piston area, and S is the stroke of the piston. In conventional fluid power piston pumps, the several pistons are arranged radially around a shaft, as shown in Fig. 18–6. Thus the displacement of such a pump would be the product of the displacements of each cylinder and the number of cylinders in the pump. Because of the large number of variables involved and the number of different designs in use, there is no simple formula for the output of a piston pump.

Positive displacement fluid power pumps are available in pressure ratings up to 10,000 psi. Some are capable of delivering 100–150 gpm. For most applications, horsepower capability is currently in the range of 5 to 1000 hp. Components providing higher horsepower are available.

Since PD pumps are mechanical devices, their design must take into account clearances between mating mechanical parts, such as the gears and housing in a gear pump, the pistons and cylinder bores in a piston pump, etc. These clearances provide small but definite paths within the pumps for the fluid to leak back from the pressure or discharge side of the pump to the suction or inlet side of the pump. Such *internal leakage* in a PD pump is called *slip* leakage. Its existence means that less fluid is actually delivered to the external circuit than would be expected from the theoretical displacement calculated by Eqs. (18–1) or (18–2). This loss in pumping capability due to internal leakage is reflected by the *volumetric efficiency,* which is the ratio of the actual output from the pump to the theoretical output, expressed in percent:

$$e_v = (Q_a/Q_t) \times 100 = (V_a/V_d) \times 100. \tag{18–3}$$

Volumetric efficiency is always less than 100%. For commercial gear pumps it ranges from 70 to 90%, depending upon the quality and design of the pump. For vane pumps, it is $80 < e_v < 92\%$, again depending upon the design of the unit, quality, and operating pressures. For piston pumps, which have the highest performance of the three types, volumetric efficiency should lie between 90 and 98%.

Positive displacement pumps are also subject to mechanical losses, and therefore a *mechanical efficiency* factor is used to express the difference between the theoretical input and the actual input torques:

$$e_m = \frac{\text{theoretical input torque} \times 100\%}{\text{actual input torque}}. \tag{18–4}$$

This holds true for any mechanical device; e_m is always less than 100%.

The product of the volumetric efficiency and the mechanical efficiency is called the *overall efficiency,* and is used to describe the relative performance of a positive displacement pump as an energy transfer device in a fluid power system.

Energy Transfer Devices, Output

Motors

Just as the pump performed the function of energy transfer to the fluid system for transmission to some remote point, fluid motors extract the energy from the fluid and convert it to a mechanical output to do some useful work. The process is the direct opposite of the pumping process.

There are three basic classes of fluid power motors. The first is related to linear motors. The other two are rotary motors.

1) the hydraulic or pneumatic cylinders, with which we have dealt in earlier chapters on pressure, area, and force relationships.

2) Limited rotation motors, which can be called rotary cylinders.

3) Continuous rotation motors, which are similar in design and construction to the types of pumps just discussed, i.e., gear, vane, and piston.

The above types of output devices have been discussed in detail in Chapters 2, 4, and 17. Rotary fluid motors that have the same design as pumps have similar characteristics to those of pumps, except that pressurized fluid enters the motor as spent fluid exits and is returned to the pump for a new charge of energy.

Energy control devices

During the course of our study of basic fluid mechanics, we have seen that the level of energy transfer in a fluid power system is determined by the pressure which we observe. We also know that energy is transferred via changes in potential, rather than kinetic, energy in a fluid power system. Furthermore, we know that the rate of energy transfer is dependent upon the flow rate of fluid in the system.

If we now add the one additional function of switching the flow, or directing the fluid from the pump to a selected location in the fluid power circuit, we have the three basic modes of control which we wish to exercise:

1) Direction control,

2) Pressure control,

3) Flow control.

Direction control valves

The fluid power components which perform the flow-switching function are called *direction control valves.* This is an unfortunate choice of nomenclature, since the valves are actually fluid switches; but the term is the current usage in industry.

In fluid power, particularly in the case of hydraulic systems, we are dealing with an energy transfer medium, i.e., a fluid which has definite volume and mass. Thus any switching or control valve devices which we use must have a provision for the passage of this relatively massive fluid medium. Figure 18–8 illustrates the basic methods of achieving valving action in fluid control devices. Figure 18–8(a) shows that flow can be turned on or off by plugging a hole, or port, in the valve. A component of this type is called a *seating* valve element. Figure 18–8(b) shows that another method of achieving control is by sliding something over a hole or port. A component of this type is a *slider or sliding* element. In Fig.

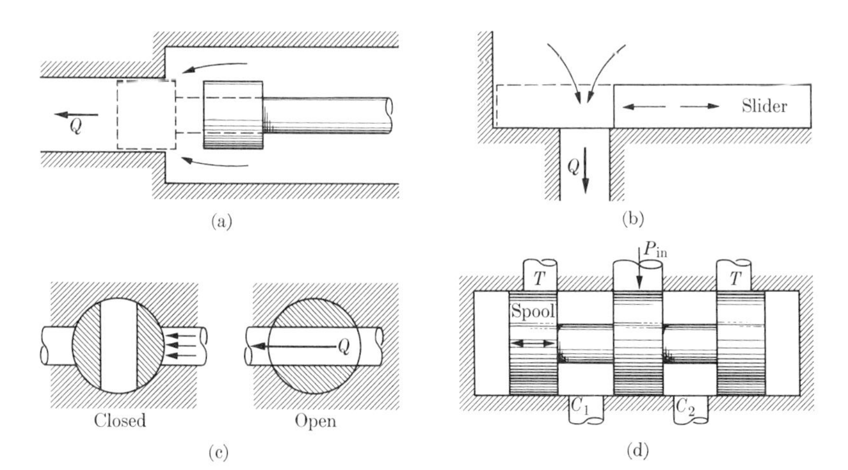

Fig. 18–8

18–8(c) we have a *rotating* element. Flow is turned on or off by rotation of the element which has a port. Figure 18–8(d) illustrates the most commonly used device, the *spool* element, which is moved back and forth in a cylindrical bore. As it moves, the sections of reduced diameter come in contact with ports in the valve body and not only can turn flow off or on but also can switch it from one circuit to another.

Although the function of the directional control devices is to switch flow, they also cause flow losses in the system, since they function in an "orifice-like" manner; that is, they introduce a restriction in the line. In Chapter 9, we discussed some of these devices in more detail. In spite of the fact that pressure drops are developed across directional control valves, their primary function is not associated with controlling the pressure level in the circuit. Figure 18–9 typifies a modern directional control valve design.

Pressure control valves

In Chapters 8 and 9, we discussed orifices and the orifice-like devices. We recall the general form of the equation describing flow through an orifice is

$$Q = C_d A \sqrt{2gh},$$

which shows us that the flow rate is a function of the area of the orifice opening, the head differential across the orifice, the acceleration due to gravity, and a characteristic discharge coefficient. We can rewrite this equation in the form:

$$Q = K_o \sqrt{\Delta P},$$

Fig. 18–9

Modern practice in hydraulic direction control valve technology is evident in this solenoid-operated, pilot-actuated spool valve. This unit, designed for mobile equipment applications, is rated for pressures up to 3000 psi. The design and performance of such a valve are extensions of the principles outlined in this text.

where K_o is a constant incorporating all the characteristics of a specific orifice and ΔP is the pressure drop across the orifice. Now we can see that the flow rate through a given orifice is a function of the design characteristics of that orifice (or orifice-like device) and the square root of the pressure drop. This suggests how we might design valves which would control the level of pressure in a fluid power circuit.

In most fluid power systems, we consider flow to be constant under a given set of conditions; that is, the pump is capable of delivering a constant flow rate Q. Thus if our primary interest is to control pressure, we can use an orifice-like device which will give us a ΔP as a function of the flow rate through it. Note that ΔP will vary as the square of the flow rate.

This is the technique used. There are four basic pressure control valves:

1) Relief valves,
2) Pressure reducing valves,
3) Sequence valves,
4) Unloading valves.

A relief valve is used to limit the maximum pressure which will be imposed on a system. A reducing valve controls the pressure level in a circuit by introducing a controlled pressure drop into the circuit. A sequence valve is really a fluid switching device which operates on a pressure signal. It switches flow to a secondary circuit when pressure in the primary circuit reaches a certain preset level. An unloading valve "unloads the pump," that is, bypasses its flow back to the reservoir (tank). All these classes of valves work on the orifice principle previously discussed or are switched when a certain pressure level is reached.

Flow control valves

From our previous discussion on orifice flow-pressure drop relationships, we know that control over flow can be exercised as a function of pressure drop. Thus the function of flow control valves is the reverse of that of pressure control. In the latter, we used a fixed flow rate to give a pressure drop, and thus we exercised control pressure in the system downstream from the valve. In the flow control case, we use a variable pressure drop across orifice-like devices to control the flow rate. There are many variations of this basic concept which are beyond the scope of this discussion. The basic principle underlying them all is a form of $Q = C_d A \sqrt{2gh}$.

Fluid power systems as energy transfer systems

We have discussed in great detail the concept that fluid power is the technology dealing with the transfer of energy by means of a fluid. From the point of view of the theoretical worker in fluid mechanics, the application of the techniques of fluid mechanics is subordinate to the techniques themselves. However, to the engineer the application is of greater concern.

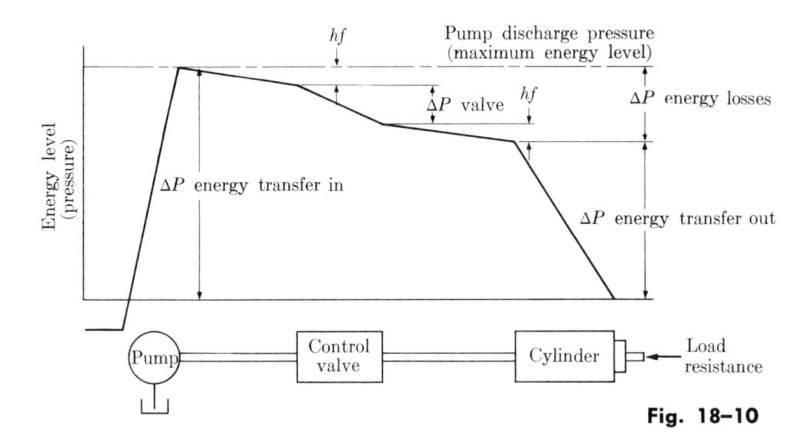

Fig. 18–10

This is particularly true if one is an advocate of the definition which holds that engineering consists of the *solution* of problems by means of the application of scientific knowledge. Thus the scientific knowledge itself, or the generation thereof, does not constitute engineering. Therefore, to the fluid power engineer and engineering technician, the application of the techniques of fluid mechanics to the transmission of energy is of prime importance.

Figure 18–10 illustrates schematically the concept of energy transfer in a fluid power system. Fluid enters the pump at a subatmospheric pressure, called the inlet or suction pressure. As it passes through the PD pump, its potential energy level is raised by means of energy transfer from the prime mover. This increase in the potential energy level is evidenced by an increase in fluid pressure. The fluid flows through the piping system and undergoes line losses; this we discussed in Chapters 14, 15, and 16. Thus it has lost some of its energy by the time it enters the control valve. In flowing through the control valve, which is an orifice-like device, the fluid undergoes a further energy loss. This is evidenced by a pressure drop across the valve. Further line losses are incurred as the fluid flows through the pipe from the valve to the motor or the output device. At the output device, which we can consider to be a mechanical interface between the energy-charged fluid system and the mechanical load, the energy is transferred to the load; that is, useful work is done. This completes the simple cycle of energy transfer in a fluid power system.

The details describing how all these functions are performed, the analytical techniques for describing these actions, and the designs, types, and functions of the various components which are used to implement these systems constitute the body of knowledge of fluid power technology. Within this text we have discussed the fundamentals of fluid mechanics on which the technology is based.

IMPORTANT TERMS

Hydrodynamic pumps are those types of pumps which transfer energy to a fluid by means of kinetic energy.

Hydrostatic pumps are pumps which transfer energy to a fluid by means of an increase in potential energy.

Positive displacement pump is a hydrostatic pump type in which a discrete volume of fluid is forcibly ejected from the pump during each cycle of operation.

Gear pump is a positive displacement pump using two or more gears in mesh to perform the pumping action.

Vane pump is a positive displacement pump which utilizes radial, sliding vanes positioned in slots in a rotor to perform the pumping function.

Piston pump is a class of positive displacement pump which uses cylindrical pistons reciprocating in cylinder bores to provide the pumping action.

Volumetric efficiency, expressed in percent, is the ratio of the actual displacement of a PD pump to its theoretical displacement.

Mechanical efficiency, expressed in percent, of a PD pump is the ratio of theoretical torque necessary to operate it at any given operating point to the actual torque required to run it.

Overall efficiency is the product of the volumetric and the mechanical efficiencies.

Direction control valves are fluid switching devices which route or switch fluid from one pipe line to another.

Pressure control valve is a fluid power control device used to control pressure level in a circuit or to perform some function related to pressure level.

Flow control valve is a fluid power control device which regulates flow rate in a circuit downstream from the position of the valve.

PROBLEMS

18–1 The basic factors to be considered in a pumping system are shown in Fig. 18–11. Neglecting losses, one can use Bernoulli's equation to characterize the system. The pump is driven by a 10-hp electric motor. Pump efficiency is 80%. $P_1 = -3$ psi, d_1 (pipe) $= 2$ in., d_2 (pipe) $= 1\frac{1}{2}$ in., and the suction lift is 5 ft. Assuming the motor is fully loaded, determine the height to which the water can be raised.

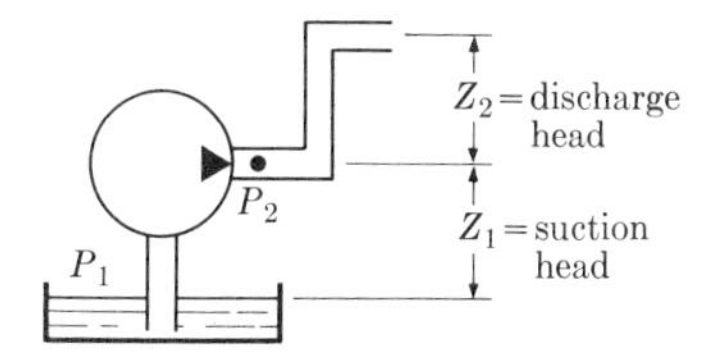

Fig. 18–11

18–2 A centrifugal water pump impeller 18 in. in diameter rotates at 1800 rpm. Assuming no flow through the pump, determine the maximum pressure developed at the discharge port.

18–3 A centrifugal pump operating at 2000 rpm develops a head of 180 ft at a flow rate of 1000 gpm. What is its specific speed?

18–4 A gear-type pump is designed with gears having an outside diameter of 3 in. and a root diameter of 2 in. The gears are 2 in. long. Calculate the displacement of a two-gear pump. At a prime mover speed of 1800 rpm, determine the flow rate. The discharge pressure is 1200 psi. Calculate the hydraulic horsepower. If the pump is 80% efficient, what size prime mover must be used?

18–5 Repeat Problem 18–4, using a three-gear pump.

18–6 A vane pump has a cam ring diameter of 6 in. Eccentricity between the cam and rotor is 0.250 in. Rotor elements are 1 in. wide. Calculate the pump displacement. Plot the curve of the flow rate versus the speed of the prime mover.

18–7 A piston pump has 9 pistons of $\frac{1}{2}$-in. diameter arranged axially around the shaft center. The stroke is a maximum of $\frac{1}{2}$ in. Neglecting losses, calculate the displacement. What is the flow rate at 1200 rpm? If discharge pressure is 3000 psi, what is the output horsepower? If the pump is 91% efficient, determine the horsepower of the prime mover.

18–8 Consider the gear pump of Problem 18–4. If the volumetric efficiency is 88% and the mechanical efficiency is 85%, repeat the calculations of 18–4. At the 1800-rpm speed, what is the actual leakage or slip rate?

18–9 If the volumetric efficiency in the vane pump of Problem 18–6, is 90% and the mechanical efficiency is 95%, repeat the calculations. N = 1800 rpm; P = 1500 psi.

18–10 If the volumetric efficiency is 95% and the mechanical efficiency is 93%, recalculate the data for the piston pump of Problem 18–7.

18–11 If the pistons and cylinders of Problem 18–7 are placed on a 6-in. centerline circle and mineral-based hydraulic oil is the fluid, what is the pressure generated by centrifugal force due to the rotation at 1200 rpm? (See Chapter 12.) What effect would this pressure have on the flow of oil into the pump during the suction stroke?

18–12 What would be the centrifugal pressure force in the gear pump of Problem 18–4 and the vane pump of Problem 18–6?

18–13 Consider a gear motor of the same design as the pump of Problem 18–1. If oil enters the motor at 1500 psi at a rate of 60 gpm and if we neglect losses, what would be the output speed of the motor? If it were actually 1700 rpm, what would be the volumetric efficiency of the motor?

18–14 Under the conditions of Problem 18–13, what was the horsepower delivered to the motor? What is the motor's output torque? What was the actual output horsepower? Calculate the motor's efficiency.

18–15 What are the basic methods for achieving valving action in direction control valves?

18–16 What is the control technique common to both pressure and flow control valves? Explain how this same technique is used to achieve two different control functions.

18–17 What is the fundamental difference between directional control and pressure or flow control valves?

18–18 Discuss the concept of a "fluid power system as an energy transfer system" as you see it, and relate the concept to the fundamentals of fluid mechanics which we have discussed in this text.

Appendix A

Derivation of Equations for Calculating Hydrostatic Force on a Submerged Surface

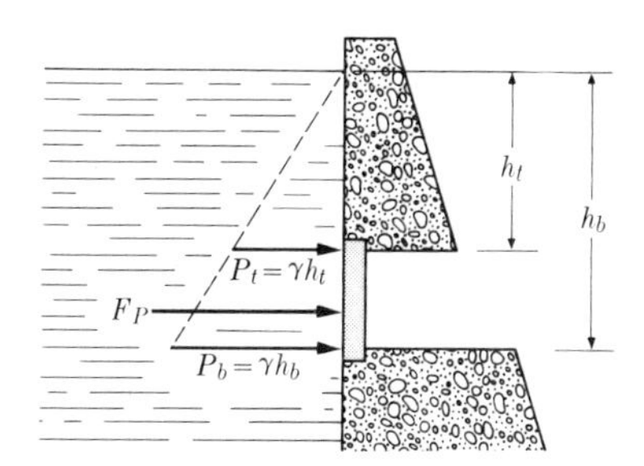

Fig. A-1

Let us suppose that the submerged plate shown in Fig. A-1 is actually a gate in a dam. What will be the total pressure force, F_P, acting on it? Where will this resultant act? From

$$F = P \cdot A, \qquad dF = P\,dA,$$

we have

$$F = \int P\,dA, \qquad dA = w \cdot dh, \qquad P = \gamma h,$$

$$F = \int_{h_t}^{h_b} \gamma h w\,dh,$$

$$F = \gamma w \int_{h_t}^{h_b} h\,dh = \gamma w \left.\frac{h^2}{2}\right|_{h_t}^{h_b} = \frac{\gamma w}{2}\,(h_b^2 - h_t^2).$$

Note that

$$\text{Area} = w(h_b - h_t),$$

$$\bar{h} = \frac{h_b + h_t}{2}\ \overline{P} = \gamma\bar{h}.$$

Thus

$$F = \gamma \cdot \frac{(h_b + h_t)}{2} \cdot w(h_b - h_t) = \overline{P} \cdot A,$$

where $\overline{P}$ is the pressure at the centroid of the plate and A is the area of the plate.

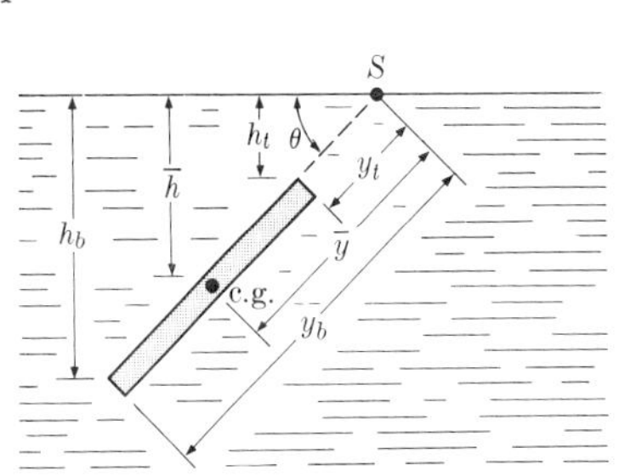

Fig. A–2

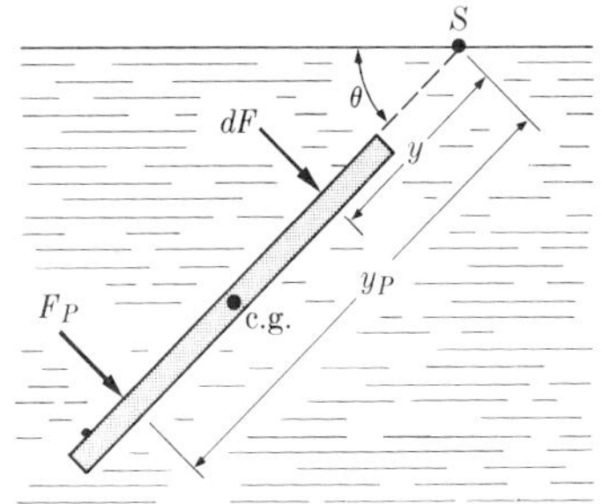

Fig. A–3

Next let us suppose that the plate is not vertical (Fig. A–2). Extend the plane of the plate until it intersects the surface at S. The heads will then be measured as the sin θ function of the distance y measured along the plane of the plate. Then

$$F = \int_{y_t}^{y_b} \gamma \sin \theta y \, dA,$$

$$F_p = \gamma\bar{h}A = \overline{P} \cdot A = \text{resultant pressure force.}$$

[*Rule:* The total force on a submerged plane surface is the product of the pressure at its centroid and the area of the surface.]

To determine the point of application of the resultant pressure force on a submerged plane surface, consider the following:

a) Do not confuse the *total force,* as calculated above, with the *point of application* of the total force. Take the moment of forces about the surface. (See Fig. A–3.)

b) $M_S = \int y \, dF$. Also $M_S = F_P \cdot y_P$. Then $F_P \cdot y_P = \int y \, dF$. We have just shown that $dF = \gamma \sin \theta y \, dA$. Then

$$F_P \cdot y_P = \int \gamma \sin \theta y^2 \, dA,$$

and hence

$$y_P = \frac{\int \gamma \sin \theta y^2 \, dA}{F_P} = \frac{\int y^2 \, dA}{\int y \, dA},$$

but

$$I = \int y^2 \, dA, \qquad \bar{y} = \text{distance to center of gravity (c.g.)}.$$

So

$$y_P = \frac{I}{A \cdot \bar{y}}.$$

The moment of inertia of the plane area about the axis through the point S can be written in the form

$$I = I_{\text{c.g.}} + A \cdot \bar{y}^2,$$

where $I_{\text{c.g.}}$ is the moment of inertia about the axis through c.g. Then

$$y_P = \frac{I_{\text{c.g.}} + A \cdot \bar{y}^2}{A \cdot \bar{y}} = \frac{I_{\text{c.g.}}}{A \cdot \bar{y}} + \frac{A \cdot \bar{y}^2}{A \cdot \bar{y}},$$

since

$$\frac{A \cdot \bar{y}^2}{A \cdot \bar{y}} = \bar{y}, \qquad \text{and} \qquad y_p - \bar{y} = \frac{I_{\text{c.g.}}}{A \cdot \bar{y}},$$

which shows that the distance from the centroid to the center of pressure $(y_P - \bar{y})$ equals the moment of inertia about c.g. divided by the moment of the area about S. (See Fig. A–4.)

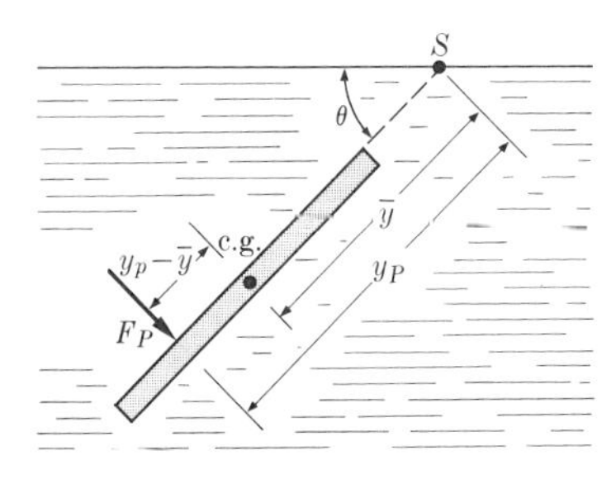

Fig. A–4

The radius of gyration of a plane area $= \sqrt{I/A} = K$ such that $I_{\text{c.g.}} = K^2 \cdot A$. Then

$$y_P - \bar{y} = \frac{K^2 \cdot A}{A \cdot \bar{y}} = \frac{K^2}{\bar{y}} = \frac{I_{\text{c.g.}}}{A \cdot \bar{y}} \equiv \frac{K^2}{\bar{y}}.$$

Appendix B

TABLE B–1 SOME PROPERTIES OF TYPICAL LIQUIDS

Liquid	*Sg*	*Density slug/ft³*	*Absolute viscosity lb-sec/ft²*	*Kinematic viscosity ft²/sec*
		General fluids		
Ethyl alcohol	.765	1.53	3×10^{-5} 50°F	1.96×10^{-5} 50°F
Benzene	.853	1.71	1.6×10^{-5} 50°F	9.35×10^{-6} 50°F
Carbon tetrachloride	1.545	3.09	2.4×10^{-5} 50°F	7.77×10^{-6} 50°F
Castor oil	.97 20°C	1.83 30°C	6×10^{-3} 100°F	3.18×10^{-3} 100°F
Gasoline	.68	1.32	7×10^{-6} 50°F	5.3×10^{-6} 50°F
Glycerine	1.22	2.45	4×10^{-3} 100°F	1.63×10^{-3} 100°F
Kerosene	.78	1.52	5×10^{-5} 50°F	3.31×10^{-5}
Mercury	13.6	26.3	3.5×10^{-5} 50°F	1.33×10^{-6} 50°F
Water, potable	1.0 39.2°F	1.94	2.6×10^{-5} 50°F	1.34×10^{-5} 50°F
Water, sea	1.03 15°F	1.99	—	—

TABLE B–1 *(Continued)*

Liquid	*Sg*	*Density slug/ft³*	*Absolute viscosity lb-sec/ft²*	*Kinematic viscosity ft²/sec*
		Hydraulic fluids		
Mineral oil	0.89	1.73	8.20×10^{-4}	4.73×10^{-4} 100°F
SAE 10	0.90	1.75	8.20×10^{-4}	4.73×10^{-4} 100°F
SAE 30	0.90	1.75	1.89×10^{-3}	1.08×10^{-3} 100°F
MIL 5606	0.86	1.67	2.69×10^{-4}	1.62×10^{-4} 100°F

Mineral Oil: assume 200 SSU $= 44 \text{ CST} = \dfrac{44}{9.3000} = 4.73 \times 10^{-4} \dfrac{\text{ft}^2}{\text{sec}}$

TABLE B–2 PROPERTIES OF SOME GASES

Gas	Specific weight, w, at 68°F, 1 atm, lb/ft^3	Gas constant, R, ft/°R	Adiabatic exponent, k	Kinematic viscosity, ν, at 68°F, 1 atm, ft^2/sec
Air	0.0752	53.3	1.40	16.0×10^{-5}
Ammonia	0.0448	89.5	1.32	16.5
Carbon dioxide	0.1146	34.9	1.30	9.1
Methane	0.0416	96.3	1.32	19.3
Nitrogen	0.0726	55.1	1.40	17.1
Oxygen	0.0830	48.3	1.40	17.1
Sulfur dioxide	0.1695	23.6	1.26	5.6

TABLE B–3 PROPERTIES OF AIR AT ATMOSPHERIC PRESSURE AT DIFFERENT TEMPERATURES

Temperature, °F	Density, ρ, slug/ft^3	Specific weight, w, lb/ft^3	Kinematic viscosity, ν, ft^2/sec	Dynamic viscosity, μ, lb-sec/ft^2
0	0.00268	0.0862	12.6×10^{-5}	3.28×10^{-7}
20	0.00257	0.0827	13.6	3.50
40	0.00247	0.0794	14.6	3.62
60	0.00237	0.0763	15.8	3.74
68	0.00233	0.0752	16.0	3.75
80	0.00228	0.0735	16.9	3.85
100	0.00220	0.0709	18.0	3.96
120	0.00215	0.0684	18.9	4.07

TABLE B–4 FRICTIONAL FACTORS f FOR WATER ONLY

(Temperature range about 50°F to 70°F)

For old pipe — approximate range ϵ from 0.004 ft to 0.020 ft
For average pipe — approximate range ϵ from 0.002 ft to 0.003 ft
For new pipe — approximate range ϵ from 0.0005 ft to 0.0010 ft

(f = tabular value $\times 10^{-4}$)

Diameter, in.	*Type of pipe*	*Velocity, ft/sec*										
		1	2	3	4	5	6	8	10	15	20	30
4	*Old, comm.*	435	415	410	405	400	395	395	390	385	375	370
	Average, comm.	355	320	310	300	290	285	280	270	260	250	250
	New pipe	300	265	250	240	230	225	220	210	200	190	185
	Very smooth	240	205	190	180	170	165	155	150	140	130	120
6	*Old, comm.*	425	410	405	400	395	395	390	385	380	375	365
	Average, comm.	335	310	300	285	280	275	265	260	250	240	235
	New pipe	275	250	240	225	220	210	205	200	190	180	175
	Very smooth	220	190	175	165	160	150	145	140	130	120	115
8	*Old, comm.*	420	405	400	395	390	385	380	375	370	365	360
	Average, comm.	320	300	285	280	270	265	260	250	240	235	225
	New pipe	265	240	225	220	210	205	200	190	185	175	170
	Very smooth	205	180	165	155	150	140	135	130	120	115	110
10	*Old, comm.*	415	405	400	395	390	385	380	375	370	365	360
	Average, comm.	315	295	280	270	265	260	255	245	240	230	225
	New pipe	260	230	220	210	205	200	190	185	180	170	165
	Very smooth	200	170	160	150	145	135	130	125	115	110	105
12	*Old, comm.*	415	400	395	395	390	385	380	375	365	360	355
	Average, comm.	310	285	275	265	260	255	250	240	235	225	220
	New pipe	250	225	210	205	200	195	190	180	175	165	160
	Very smooth	190	165	150	140	140	135	125	120	115	110	105

TABLE B–5 PROPERTIES OF AMERICAN STANDARD PIPE

Nominal size, in.	Number of threads/in.	Diameter, actual external	Tap drills & diameter bore	Standard								Extra heavy							
				Diameter, actual internal	Internal area, in²	gpm at 1 ft/sec, velocity	Nominal weight, lb/ft	Bursting pressure				Diameter, actual internal	Internal area, in²	gpm at 1 ft/sec, velocity	Nominal weight, lb/ft	Bursting pressure			
								(a) Lap weld	*	(b) Butt weld	*					(a) Lap weld	*	(b) Butt weld	*
$\frac{1}{8}$	27	0.405	$\frac{11}{32}$	0.27	0.06	0.18	0.24	—	1250	13750	2500	0.22	0.04	0.11	0.31	—	2000	19800	3500
$\frac{1}{4}$	18	0.540	$\frac{7}{16}$	0.36	0.10	0.32	0.42	—		13350		0.30	0.07	0.22	0.54	—		19000	
$\frac{3}{8}$	18	0.675	$\frac{19}{32}$	0.49	0.19	0.60	0.57	—		11050	1500	0.42	0.14	0.44	0.74	—		15500	2500
$\frac{1}{2}$	14	0.840	$\frac{23}{32}$	0.62	0.30	0.95	0.85	—		10650		0.55	0.23	0.73	1.09	—		14600	
$\frac{3}{4}$	14	1.050	$\frac{15}{16}$	0.82	0.53	1.66	1.13	—		8850		0.74	0.43	1.35	1.47	—		11900	2000
1	$11\frac{1}{2}$	1.315	$1\frac{5}{32}$	1.05	0.86	2.69	1.68	—		8275		0.96	0.72	2.24	2.17	—		11400	
$1\frac{1}{4}$	$11\frac{1}{2}$	1.660	$1\frac{1}{2}$	1.38	1.50	4.46	2.27	—		6900	1000	1.28	1.28	4.00	3.00	—		9600	1500
$1\frac{1}{2}$	$11\frac{1}{2}$	1.900	$1\frac{23}{32}$	1.61	2.04	6.35	2.72	—		6275		1.50	1.77	5.51	3.63	—		8850	
2	$11\frac{1}{2}$	2.375	$2\frac{3}{16}$	2.07	3.36	10.5	3.65	6750		5325	750	1.94	2.95	9.20	5.02	9750		7700	1250
$2\frac{1}{2}$	8	2.875	$2\frac{5}{8}$	2.47	4.79	14.9	5.79	7350		5800		2.32	4.24	13.2	7.66	10100	1500	7950	
3	8	3.500	$3\frac{1}{4}$	3.07	7.39	23.0	7.58	6425	1000	5050		2.90	6.61	20.6	10.25	9150		7200	
4	8	4.500	$4\frac{1}{4}$	4.03	12.73	39.7	10.79	5475		—		3.83	11.50	35.8	14.98	7900	1250	—	
5	8	5.563	$5\frac{5}{16}$	5.05	20.01	62.4	14.62	4825	750	—		4.81	18.19	56.7	20.78	6950		—	
6	8	6.625	$6\frac{3}{8}$	6.07	28.89	90.0	18.97	4375		—		5.76	26.07	81.2	28.57	6850		—	
8	8	8.625	$8\frac{11}{32}$	7.98	50.02	156.0	28.55	3875	650	—		7.63	45.66	142	43.34	6050	1000	—	
10	8	10.75	$10\frac{7}{16}$	10.02	78.85	246.0	40.48	3525	500	—		9.75	74.66	233	54.74	4825	750	—	
12	8	12.75	$12\frac{7}{16}$	12.00	113.1	353.0	49.56	3050		—		11.75	108.4	338	65.42	4075		—	

Nominal size, in.	Number of threads/in.	Diameter, actual external	Tap drills & diameter bore	Double extra heavy: Diameter, actual internal	Internal area, in^2	gpm at 1 ft/sec, velocity	Nominal weight, lb/ft	Bursting pressure: (a) Lap weld	*	(b) Butt weld	*	Seamless tubes: Diameter, actual internal	Internal area, in^2	gpm at 1 ft/sec, velocity	Nominal weight, lb/ft	Bursting pressure	*
$\frac{1}{8}$	27	0.405	$\frac{11}{32}$	—	—	—	—	—	4000	—	5000	—	—	—	—	—	6000
$\frac{1}{4}$	18	0.540	$\frac{7}{16}$	—	—	—	—	—		—		—	—	—	—	—	
$\frac{3}{8}$	18	0.675	$\frac{19}{32}$	—	—	—	—	—		—		—	—	—	—	—	
$\frac{1}{2}$	14	0.840	$\frac{23}{32}$	0.25	0.05	0.16	1.71	—		28900		.252	0.050	0.16	1.71	43600	
$\frac{3}{4}$	14	1.050	$\frac{15}{16}$	0.43	0.15	0.46	2.44	—		24300	4000	0.434	0.148	0.46	2.44	36750	
1	$11\frac{1}{2}$	1.315	$1\frac{5}{32}$	0.60	0.28	0.88	3.66	—		22350		0.599	0.282	0.88	3.66	33750	
$1\frac{1}{4}$	$11\frac{1}{2}$	1.660	$1\frac{1}{2}$	0.90	0.63	1.96	5.21	—		19000	3000	0.896	0.630	1.96	5.21	28750	4000
$1\frac{1}{2}$	$11\frac{1}{2}$	1.900	$1\frac{23}{32}$	1.10	0.95	2.96	6.41	—		17200		1.100	0.950	2.96	6.41	26000	
2	$11\frac{1}{2}$	2.375	$2\frac{3}{16}$	1.50	1.77	5.53	9.03	19000		15000	2500	1.503	1.774	5.53	9.03	22700	
$2\frac{1}{2}$	8	2.875	$2\frac{5}{8}$	1.77	2.46	7.68	13.70	19950	3500	15750		1.771	2.464	7.63	13.70	23800	
3	8	3.500	$3\frac{1}{4}$	2.30	4.16	13.0	18.58	17800	3000	—		2.300	4.155	13.0	18.58	21200	3500
4	8	4.500	$4\frac{1}{4}$	3.15	7.80	24.3	27.54	15600	2500	—		3.152	7.803	24.3	27.54	18600	3000
5	8	5.563	$5\frac{5}{16}$	4.06	12.97	40.4	38.55	14000		—		4.063	12.966	40.4	38.55	16750	
6	8	6.625	$6\frac{3}{8}$	4.90	18.84	58.7	53.16	13700		—		4.897	18.835	58.7	53.16	16350	2500
8	8	8.625	$8\frac{11}{32}$	6.88	37.12	116.0	72.42	10600	2000	—		6.875	37.122	116.0	72.42	12650	2000
10	8	10.75	$10\frac{7}{16}$	—	—	—	—	—		—		—	—	—	—	—	
12	8	12.75	$12\frac{7}{16}$	—	—	—	—	—		—		—	—	—	—	—	

LIST OF SYMBOLS AND ABBREVIATIONS

Symbol	*Definition*	*Dimensions*	*Units*
a	Linear acceleration	Length/time^2	in/sec^2, ft/sec^2
A	Area	Length^2	in^2, ft^2
C	Various constants	Usually dimensionless, otherwise as defined	
d	Diameter (when used with capital D, it usually denotes inside or smaller diameter)	Length	in., ft, cm, m
D	Diameter, larger or outside, subscripts sometimes used	Length	in., ft, cm, m
e	Eccentricity of one centerline relative to another	Length	in., ft, cm, m
f	Friction factor	Dimensionless	see Chaps. 14, 15
F	Force	Force	pounds, ounces, dynes
g	Acceleration due to gravity	Length/time^2	in./sec^2, ft/sec^2
G	Work	Length · force	ft-lb, in-lb, dyne-cm
h or H	Head	Length	in., ft, etc.
HP	Horsepower	unit of rate of energy transfer	550 ft-lb/sec/hp, 33,000 ft-lb/min/hp
J	Polar moment of inertia	force · length · time^2	lb · ft · sec^2, lb · in. · sec^2
k or K	Various constants, usually associated with flow losses in pipe, fittings, valves, etc.	Usually dimensionless numbers	
l	Linear distance	Length	in., ft, cm, m

LIST OF SYMBOLS AND ABBREVIATIONS (*Continued*)

Symbol	*Definition*	*Dimensions*	*Units*
L_e	Equivalent length	Length	in., ft, cm, m
m or M	Mass	$\frac{\text{force} \cdot \text{time}^2}{\text{length}}$	$\text{lb} \cdot \text{sec}^2/\text{ft}$ = slug, $\text{dyne} \cdot \text{sec}^2/\text{cm}$
M_1	By definition, mass flow rate, refer to Chapter 11	$\frac{\text{force} \cdot \text{time}}{\text{length}}$	slug/sec = $\frac{\text{lb} \cdot \text{sec}}{\text{ft}}$
N_R	Reynolds number	Dimensionless	
P	Pressure (distributed reaction)	Force/length^2 (force/area)	lb/in^2, lb/ft^2, dyne/cm^2
p	Power (rate of energy transfer)	$\frac{\text{force} \cdot \text{length}}{\text{time}}$	ft-lb/sec, ft-lb/min
q or Q	Flow rate (of fluid)	*Volumetric:* $\text{length}^3/\text{time}$ gal/time	ft^3/min, in^3/sec, gpm or gps
		Weight: force/time	lb/sec, lb/min
		Mass: mass/time	$\text{lb} \cdot \text{sec/ft}$ = slug
r or R	Radius	Length	ft, in., cm, etc.
s	Distance	Length	ft, in., cm, etc.
S	Stroke (of cylinder)	Length	ft, in., cm, etc.
S_g	Specific gravity	Dimensionless	
t	Time	Time	Sec, min
T	Torque	$\text{Force} \cdot \text{length}$	$\text{lb} \cdot \text{ft}$, $\text{lb} \cdot \text{in.}$
u	Relative velocity	Length/time	in./sec, ft/sec, in./min, ft/min
v	Absolute velocity	Length/time	in./sec, ft/sec, in./min, ft/min
V	Volume	Length^3	in^3, ft^3, cc or cm^3
V_0	Displacement volume	Length^3	in^3, ft^3, cc (cm^3)

LIST OF SYMBOLS AND ABBREVIATIONS (*Continued*)

Symbol	*Definition*	*Dimensions*	*Units*
w or W	Weight (force on a mass due to gravitational acceleration)	Force	lb, oz, g, etc.
x	Generally used to denote position or length along the abscissa in a coordinate system	Depends on definition of abscissa	
y	Generally used to denote length or position along ordinate axis in a coordinate system	Depends on definition of ordinate axis	
z	Generally used to denote length or position along an axis perpendicular to the abscissa and ordinate	Depends on definition of ordinate axis	
Z	Elevation head	Length	in., ft, etc.

GREEK ALPHABET

A	α	Alpha	N	ν	Nu
B	β	Beta	Ξ	ξ	Xi
Γ	γ	Gamma	O	o	Omicron
Δ	δ	Delta	Π	π	Pi
E	ϵ	Epsilon	P	ρ	Rho
Z	ζ	Zeta	Σ	σ	Sigma
H	η	Eta	T	τ	Tau
Θ	θ	Theta	Υ	υ	Upsilon
I	ι	Iota	Φ	ϕ, φ	Phi
K	κ	Kappa	X	χ	Chi
Λ	λ	Lambda	Ψ	ψ	Psi
M	μ	Mu	Ω	ω	Omega

GREEK SYMBOLS

Letter	*Phonetic*	*Description*	*Dimensions*	*Units*
α	Alpha	Angular acceleration	Angle/time^2	Rad/sec^2
β	Beta	Bulk modulus	Force/length^2	lb/in^2
ϵ	Epsilon	Absolute height of roughness projection in fluid conductor	Length	in, ft, cm, m
γ	Gamma	Specific weight of a substance	Force/length^3	lb/in^3, lb/ft^3
δ	Delta	Increment in measurement or variable	Depends on use	———
θ	Theta	Angular displacement		degrees/rad
μ	Mu	In fluid mechanics, symbol for absolute viscosity	$\frac{\text{force} \cdot \text{time}}{\text{length}^2}$	lb-sec/ft^2, dyne-sec/cm^2
ν	Nu	In fluid mechanics, symbol for kinematic viscosity	Length^2/time	ft^2/sec, cm^2/sec
π	Pi	Ratio of circumference of a circle to its diameter	———	3.14159 + . . .
ρ	Rho	In fluid mechanics, density of a substance	$\frac{\text{force} \cdot \text{time}^2}{\text{length}}$	slugs, lb-sec^2/ft
Σ	Sigma	Summation	———	———
τ	Tau	In fluid mechanics, viscous shear stress	force/length^2	lb/in^2, dynes/cm^2
ω	Omega	Angular velocity	rad/time	rad/sec, rad/min

ABBREVIATIONS

cfs	Cubic feet per second
cis	Cubic inches per second
cfm	Cubic feet per minute
cim	Cubic inches per minute
cps	Cycles per second
cs	Centistokes (viscosity)
fps	Feet per second
fpm	Feet per minute
gpm	Gallons per minute
ips	Inches per second
ipm	Inches per minute
psi	Pounds per square inch, gage
psf	Pounds per square foot, gage
psia	Pounds per square inch, absolute
psfa	Pounds per square foot, absolute
scfm	Standard cubic feet per minute
SSU	Saybolt Seconds Universal (viscosity)
V.I.	Viscosity index

Index